How to Repair Automotive Dents, Fix Bumpers, and Eliminate Hail Damage

The Ultimate Guide to Mastering Auto Body Repair Techniques for Flawless Results

The Fix It Guy

Copyright © 2024 by The Fix It Guy

All rights reserved. No part of this book may be reproduced in any form or by any electronic or mechanical means, including information storage and retrieval systems, without permission in writing from the publisher, except by a reviewer who may quote brief passages in a review.

Table of Contents

Introduction

Chapter 1: Understanding Automotive Dents
- Types of Dents and Their Causes
- Assessing the Severity of Dents
- Determining the Best Repair Approach

Chapter 2: Paintless Dent Repair (PDR) Techniques
- Introduction to PDR
- Essential PDR Tools and Equipment
- Step-by-Step PDR Process
- Advanced PDR Techniques for Challenging Dents
- Common PDR Mistakes and How to Avoid Them

Chapter 3: Traditional Dent Repair Methods
- Filling and Sanding Techniques
- Using Body Filler for Dent Repair
- Priming and Painting Repaired Areas
- Blending New Paint with Existing Paint

Chapter 4: Bumper Repair and Replacement
- Assessing Bumper Damage
- Repairing Plastic Bumpers
- Fixing Cracked or Broken Bumpers
- Replacing Irreparable Bumpers
- Painting and Finishing Bumpers

Chapter 5: Hail Damage Repair
- Understanding Hail Damage
- Evaluating the Extent of Hail Damage

- PDR Techniques for Hail Damage Repair
- Conventional Repair Methods for Severe Hail Damage
- Insurance Claims and Hail Damage Repair

Chapter 6: Rust Repair and Prevention
- Identifying Rust Damage
- Removing Rust from Metal Surfaces
- Treating and Preventing Future Rust
- Painting and Sealing Repaired Areas

Chapter 7: Advanced Auto Body Repair Techniques
- Working with Aluminum Body Panels
- Repairing Plastic Components
- Fixing Dents on Difficult Locations
- Custom Modifications and Bodywork

Chapter 8: Troubleshooting Common Auto Body Repair Issues
- Paint Imperfections and How to Fix Them
- Uneven or Misaligned Body Panels
- Dealing with Stubborn Dents
- Preventing and Fixing Rust Recurrence

Conclusion

Introduction

Hey there, fellow car enthusiast! If you're reading this, chances are you've experienced the heart-wrenching feeling of discovering a dent, ding, or scratch on your beloved vehicle. Whether it's the result of a rogue shopping cart, a careless driver, or a sudden hailstorm, seeing your car's once-flawless exterior marred can be downright frustrating and disheartening.

But what if I told you that you have the power to restore your car's beauty and value, all with your own two hands? That's right! With the right knowledge, tools, and techniques, you can become your own auto body repair hero and say goodbye to costly trips to the body shop.

In "How to Repair Automotive Dents, Fix Bumpers, and Eliminate Hail Damage: The Ultimate Guide to Mastering Auto Body Repair Techniques for Flawless Results," I'll be your trusty guide, sharing all the secrets and skills I've learned throughout my years of experience in the auto body repair industry. Trust me, I've seen it all – from minor door dings to massive hail damage – and I'm here to help you tackle these challenges with confidence and finesse.

Throughout this book, you'll discover:

1. The essential tools and equipment you'll need to get started on your auto body repair journey, without breaking the bank.

2. Proven techniques for assessing and repairing various types of dents, including the cutting-edge paintless dent repair (PDR) method.

3. Step-by-step guidance on fixing and replacing damaged bumpers, so you can get your car looking like new again.

4. Effective strategies for dealing with the dreaded hail damage, from small dimples to extensive destruction.

5. Insider tips and tricks for achieving a flawless, professional-looking finish that will make your friends and family wonder if you secretly moonlight as a body shop pro.

But this book isn't just about fixing your car's exterior – it's about empowering you with the skills and knowledge to take control of your vehicle's appearance and value. Imagine the pride and satisfaction you'll feel when you step back and admire your handiwork, knowing that you've saved hundreds, if not thousands, of dollars in repair costs.

So, whether you're a DIY enthusiast looking to expand your automotive skillset, or simply a car owner who wants to be prepared for life's little (and sometimes big) mishaps, this book is your ultimate roadmap to mastering auto body repair. Get ready to transform your car from dented and dinged to dazzling and delightful, all while having a blast along the way!

Let's dive in and start your journey to becoming an auto body repair maestro!

Chapter 1
Understanding Automotive Dents

Types of Dents and Their Causes

Automotive dents come in various shapes, sizes, and depths, each with its own unique set of challenges when it comes to repair. To effectively tackle these imperfections, it's crucial to understand the different types of dents and the factors that cause them. In this section, we'll dive deep into the world of automotive dents and explore their characteristics and origins.

Round Dents
Round dents are one of the most common types of automotive dents. These smooth, bowl-shaped depressions are typically caused by the impact of a rounded object, such as a ball, a rock, or even a wayward elbow. The size and depth of round dents can vary greatly depending on the force of the impact and the flexibility of the metal.

Causes:
a. Hail damage: During severe weather events, hailstones can bombard your vehicle, leaving behind a smattering of round dents across the hood, roof, and trunk.
b. Fallen objects: Stray balls, acorns, or other falling objects can create round dents upon impact.
c. Car door collisions: When a car door is opened too forcefully and hits another vehicle, it can result in a round dent.

Creased Dents
Creased dents are characterized by a distinct line or ridge where the metal has been bent or folded. These dents are often more challenging to repair than round dents, as the metal has been stretched and compressed along the crease. Creased dents can range from shallow to deep and may affect a small area or span a larger portion of a body panel.

Causes:
a. Collisions with narrow objects: Impact with a narrow, rigid object, such as a pole or a bike handlebar, can create a creased dent.
b. Automotive accidents: Low-speed collisions or sideswipes can result in creased dents along the sides or corners of a vehicle.
c. Wedged objects: If a car is pressed against a narrow object, such as a curb or a parking barrier, the pressure can cause a creased dent.

Sharp Dents
Sharp dents, also known as punched dents, are characterized by a distinct, angular appearance. These dents often have a clearly defined point of impact and may be accompanied by paint damage or even a small hole in the metal. Sharp dents can be particularly challenging to repair, as the metal has been stretched and weakened at the point of impact.

Causes:
a. Vandalism: Intentional damage, such as that caused by a key or a screwdriver, can result in sharp dents.

b. Collisions with pointed objects: Impact with a sharp or pointed object, like a broken tree branch or a metal corner, can create a sharp dent.

c. Projectiles: Small, fast-moving objects, such as stones kicked up by a passing vehicle, can cause sharp dents upon impact.

Combination Dents

Combination dents are a mix of two or more types of dents, often resulting from complex or multiple impacts. These dents can be particularly challenging to assess and repair, as each type of damage requires a different approach and technique.

Causes:

a. Multi-impact collisions: If a vehicle is involved in an accident that results in multiple impacts, combination dents may occur.

b. Sliding or scraping: If a car slides along a rough surface or scrapes against another object, it can result in a combination of creased, round, and sharp dents.

c. Unique circumstances: Unusual situations, such as a vehicle rolling over or being hit by a large, irregularly shaped object, can create combination dents.

By understanding the different types of automotive dents and their causes, you'll be better equipped to assess the damage on your vehicle and determine the most appropriate repair approach. In the following sections, we'll explore how to evaluate the severity of dents and discuss various repair techniques to help you achieve a flawless finish.

Assessing the Severity of Dents

Before diving into the repair process, it's essential to evaluate the severity of the dents on your vehicle. By assessing the extent of the damage, you can determine the most appropriate repair method and estimate the time, effort, and resources required to achieve a flawless result. In this section, we'll explore the key factors to consider when assessing the severity of automotive dents.

Depth of the Dent
The depth of a dent is one of the most crucial factors in determining its severity. Shallow dents, which are typically less than half an inch deep, are often easier to repair using paintless dent repair (PDR) techniques. These dents have not stretched the metal excessively, allowing it to be massaged back into its original shape without compromising the paint.

On the other hand, deeper dents, which extend more than half an inch into the panel, may require more extensive repair methods. These dents have likely stretched the metal beyond its elastic limit, making it difficult to restore without the use of fillers or paint.

To assess the depth of a dent, you can use a straight edge or a dent depth gauge. Place the straight edge across the dent and observe how much space exists between the bottom of the dent and the straight edge. A dent depth gauge can provide a more precise measurement, helping you determine the best course of action.

Size and Location of the Dent
The size and location of a dent can also impact its severity and the chosen repair method. Smaller dents, typically less than two inches in diameter, are often easier to repair using PDR techniques. However, larger dents may require more extensive work, particularly if they span across multiple body panels or involve complex curves and contours.

The location of the dent is another important consideration. Dents on flat, easily accessible panels, such as the hood, roof, or doors, are generally easier to repair. However, dents located on body lines, edges, or near panel boundaries can be more challenging, as they may affect the structural integrity of the vehicle or require additional disassembly.

Paint Damage
Assessing the condition of the paint is crucial when evaluating the severity of a dent. In some cases, the impact that caused the dent may have also cracked, chipped, or scratched the paint. If the paint damage is minimal and confined to the clear coat, it may be possible to repair the dent using PDR techniques and then touch up the paint.

However, if the paint damage is more extensive, involving the base coat or primer, a more comprehensive repair approach may be necessary. In these cases, the dent will need to be filled, sanded, and repainted to achieve a seamless finish.

To assess paint damage, carefully inspect the dent under bright light and look for any signs of cracking, flaking, or exposed metal. If you're unsure about the extent of the paint damage, it's always best to consult with a professional or err on the side of caution to avoid further damaging the paint during the repair process.

Access to the Dent

The accessibility of a dent is another factor to consider when assessing its severity. Some dents may be located in areas that are difficult to reach, such as the interior side of a body panel or near complex electrical or mechanical components. These dents may require extensive disassembly or specialized tools to repair properly.

In some cases, limited access to a dent may necessitate the use of alternative repair methods or even the replacement of the affected body panel. By evaluating the accessibility of a dent early in the assessment process, you can make informed decisions about the repair approach and avoid potential complications down the line.

Number of Dents

Finally, the number of dents on your vehicle can impact the overall severity of the damage and the time and resources required for repair. A single, isolated dent may be relatively straightforward to address, while multiple dents spread across various panels can be more challenging and time-consuming.

When assessing the severity of multiple dents, consider the location, depth, size, and proximity of each dent to one another. Dents that are clustered closely together may require a more comprehensive repair approach, as the surrounding metal may have been stretched or compromised.

By thoroughly assessing the severity of the dents on your vehicle, you'll be better prepared to select the most appropriate repair method and tackle the damage with confidence. In the following section, we'll explore the various repair techniques available and help you determine the best approach for your specific situation.

Determining the Best Repair Approach

Once you've assessed the severity of the dents on your vehicle, the next step is to determine the most appropriate repair approach. Several factors, such as the type of dent, its location, and the extent of the damage, will influence your decision. In this section, we'll explore the main repair methods available and provide guidance on choosing the best approach for your specific situation.

Paintless Dent Repair (PDR)
Paintless Dent Repair (PDR) is a popular and efficient method for fixing shallow dents and dings without the need for fillers or paint. PDR involves carefully massaging the dent from the backside of the panel using specialized tools, gradually coaxing the metal back into its original shape.

PDR is often the best approach when:
a. The dent is shallow (typically less than half an inch deep)
b. The paint is not cracked, chipped, or scratched
c. The dent is located on a flat or gently curved panel
d. The backside of the dent is accessible

Advantages of PDR:
- Preserves the vehicle's original paint and finish
- Faster and more cost-effective than traditional repair methods
- Environmentally friendly, as it doesn't require the use of fillers or paint

Traditional Dent Repair
Traditional dent repair involves filling the dent with a body filler material, sanding it smooth, and then repainting the affected area. This method is typically used when the dent is too deep or extensive for PDR, or when the paint has been damaged.

Traditional dent repair is often the best approach when:
a. The dent is deep (more than half an inch)
b. The paint is cracked, chipped, or scratched
c. The dent is located on a body line, edge, or near a panel boundary
d. The metal has been stretched or compromised

Advantages of traditional dent repair:
- Suitable for more severe dents and damage
- Provides a smooth, flawless finish when done properly
- Can restore the vehicle's appearance to pre-damage condition

Hybrid Repair Approach
In some cases, a combination of PDR and traditional dent repair techniques may be necessary to achieve the best results. This hybrid approach involves using PDR to remove as much of the dent as possible, then applying a minimal amount of filler to smooth out any remaining imperfections before repainting the area.
A hybrid repair approach is often the best choice when:
a. The dent is moderately deep or extensive
b. The paint damage is minimal or localized
c. The goal is to minimize the use of fillers while still achieving a flawless finish

Advantages of a hybrid repair approach:
- Combines the benefits of PDR and traditional repair methods
- Minimizes the use of fillers, preserving more of the vehicle's original structure
- Can be more cost-effective and time-efficient than a full traditional repair

Panel Replacement
In some cases, the extent of the damage may be so severe that repair is not feasible or cost-effective. If a dent is extremely large, deep, or has caused significant structural damage to the panel, replacement may be the best option.

Panel replacement is often the best approach when:
a. The dent is very deep or extensive, compromising the structural integrity of the panel
b. The damage spans multiple body panels or involves complex curves and contours
c. The cost of repair exceeds the cost of replacement
d. The panel is readily available and easy to source

Advantages of panel replacement:
- Provides a brand-new, undamaged panel
- Ensures structural integrity and proper fit
- Can be more cost-effective than extensive repairs on severely damaged panels

Professional Consultation

If you're unsure about the best repair approach for your specific situation, it's always a good idea to consult with a professional auto body technician. They can assess the damage, provide expert advice, and help you make an informed decision based on your budget, goals, and the extent of the damage.

When consulting with a professional, be sure to:
a. Provide detailed information about the dent, including its location, size, and cause
b. Share any concerns you have about the repair process or outcome
c. Ask about the technician's experience, qualifications, and recommended repair approach
d. Inquire about the estimated cost and timeline for the repair

By carefully considering the various repair approaches available and seeking professional advice when needed, you'll be well-equipped to make the best decision for your vehicle and achieve a flawless, long-lasting result.

Chapter 2
Paintless Dent Repair (PDR) Techniques

Introduction to PDR

Paintless Dent Repair (PDR) is a highly skilled and innovative technique used to remove minor dents, dings, and creases from a vehicle's body without the need for traditional filling, sanding, and repainting. This method has gained popularity in recent years due to its efficiency, cost-effectiveness, and ability to preserve the vehicle's original paint and finish.

PDR works by meticulously massaging the damaged metal back into its original shape from the backside of the panel, using specialized tools and techniques. The process relies on the principle that modern automotive paint is flexible enough to withstand the gentle manipulations required to remove dents, as long as the paint itself has not been cracked or damaged.

History and Development of PDR

The origins of paintless dent repair can be traced back to the early 1900s, when automotive manufacturers first began using steel panels for car bodies. As vehicles became more commonplace, the need for efficient and cost-effective dent repair methods grew.

In the 1930s, the first rudimentary PDR techniques were developed, using simple tools like wooden handles and metal rods to push out dents from behind the panel.

These early methods were time-consuming and often yielded less-than-perfect results.

As automotive technology advanced and paint finishes became more durable, PDR techniques evolved to keep pace. In the 1980s and 1990s, specialized tools and techniques were developed, allowing technicians to remove dents more efficiently and with greater precision.

Today, PDR is recognized as a highly skilled trade, with technicians undergoing extensive training and certification to master the art of dent removal. Advancements in technology, such as LED lighting and high-resolution cameras, have further enhanced the accuracy and effectiveness of PDR techniques.

Advantages of PDR
Paintless dent repair offers numerous benefits over traditional dent repair methods, making it an attractive option for vehicle owners and automotive professionals alike.

a. Preserves original paint: PDR allows technicians to remove dents without disturbing the vehicle's original paint and finish, eliminating the need for fillers, sanding, and repainting. This helps maintain the vehicle's appearance and value.

b. Time-efficient: Compared to traditional dent repair methods, PDR is significantly faster. Most minor dents can be removed in a matter of hours, rather than the days or weeks required for filling, sanding, and repainting.

c. Cost-effective: Because PDR eliminates the need for expensive materials and labor-intensive processes, it is often more affordable than traditional dent repair methods. This can result in substantial savings for vehicle owners.

d. Environmentally friendly: PDR does not require the use of fillers, paints, or solvents, making it a more eco-friendly option compared to traditional repair methods. This reduces the environmental impact of the repair process and helps promote sustainable practices in the automotive industry.

Limitations of PDR
While paintless dent repair is a highly effective technique, it does have some limitations. Understanding these limitations is crucial for determining whether PDR is the best approach for a specific dent or damage.

a. Depth of the dent: PDR is most effective for shallow dents, typically those that are less than half an inch deep. Deeper dents may require traditional repair methods or a hybrid approach combining PDR and filling techniques.

b. Paint damage: If the dent has caused the paint to crack, chip, or scratch, PDR alone may not be sufficient to restore the panel to its original condition. In these cases, touch-up paint or traditional repair methods may be necessary.

c. Location of the dent: Dents located on body lines, edges, or near panel boundaries can be more challenging to repair using PDR, as the metal may be harder to access or manipulate without causing further damage.

d. Stretched metal: If the impact that caused the dent has stretched the metal beyond its elastic limit, PDR may not be able to fully restore the panel to its original shape. In these cases, a combination of PDR and traditional repair methods may be required.

By understanding the principles, advantages, and limitations of paintless dent repair, you can make informed decisions about the best approach for repairing dents on your vehicle. In the following sections, we'll delve into the specific tools, techniques, and processes involved in successful PDR.

Essential PDR Tools and Equipment

To successfully perform paintless dent repair, technicians rely on a variety of specialized tools and equipment. These tools are designed to access, assess, and manipulate dents from the backside of the panel, allowing for precise and efficient removal. In this section, we'll explore the essential PDR tools and equipment, their functions, and their role in the dent repair process.

Dent Lights
Dent lights are one of the most crucial tools in a PDR technician's arsenal. These high-intensity, adjustable LED lamps are used to illuminate the dent from various angles, creating a shadow that reveals the contours and depth of the damage. By manipulating the light source, technicians can accurately assess the dent and monitor progress throughout the repair process.

Types of dent lights:
a. Fluorescent dent lights: These traditional dent lights use fluorescent bulbs to provide bright, even illumination. They are affordable and reliable but may generate more heat than LED alternatives.

b. LED dent lights: Modern LED dent lights offer superior brightness, adjustability, and energy efficiency compared to fluorescent models. They produce less heat and can be easily repositioned to optimize the illumination of the dent.

Dent Pushing Tools
Dent pushing tools are the primary instruments used to apply pressure to the backside of the dent, gradually coaxing the metal back into its original shape. These tools come in a variety of shapes, sizes, and materials to accommodate different dent locations, depths, and panel access points.

Common dent pushing tools:
a. Metal rods: Solid metal rods, typically made of steel or aluminum, are used to apply direct pressure to the dent. They come in various lengths and diameters to suit different dent sizes and locations.

b. Whale tails: Named for their distinctive shape, whale tails are flat, flexible metal tools that can be inserted between panels or into tight spaces to access hard-to-reach dents.

c. Plastic and nylon pushers: These non-marring tools are used to apply pressure to the dent without risking damage to the paint or surrounding metal. They are particularly useful for shallow dents or for finishing touches.

Glue Pulling Systems
Glue pulling systems are an alternative method of dent removal that involves attaching specialized tabs or "buttons" to the surface of the dent using hot glue. Once the glue has cooled and set, the tabs are pulled using a slide hammer or bridge puller, gently lifting the dent out from the front side of the panel.

Components of a glue pulling system:
a. Glue tabs: These small, circular or rectangular tabs are made of plastic or metal and feature a smooth, flat surface that adheres to the dent.

b. Hot glue gun: A specialized hot glue gun is used to apply a strong, heat-sensitive adhesive to the tabs, bonding them securely to the surface of the dent.

c. Slide hammer: A slide hammer is a weighted, sliding handle that attaches to the glue tab and provides the necessary force to pull the dent out.

d. Bridge puller: A bridge puller is a more advanced glue pulling tool that uses a bridge-like frame to distribute the pulling force evenly across multiple glue tabs, allowing for more controlled and precise dent removal.

Dent Taps and Knockdown Tools
Dent taps and knockdown tools are used to finesse the repair and level out any high spots or irregularities that may remain after the initial dent removal process. These tools are designed to gently tap down raised areas, creating a smooth, even surface.

Types of dent taps and knockdown tools:
a. Slapper files: These flexible, metal tools feature a curved blade that can be gently "slapped" against the raised portion of the dent to flatten it out.

b. Blending hammers: These small, lightweight hammers have a soft, non-marring face that can be used to gently tap down high spots and blend the repair area with the surrounding panel.

Accessories and Miscellaneous Tools
In addition to the primary dent removal tools, PDR technicians rely on a variety of accessories and miscellaneous items to facilitate the repair process and ensure optimal results.

Essential accessories and miscellaneous tools:
a. High-resolution cameras: Cameras are used to document the damage before, during, and after the repair process, providing a visual record of the technician's work and progress.

b. Magnets: Small, powerful magnets can be used to hold tools or lighting fixtures in place, freeing up the technician's hands for more precise manipulation of the dent.

c. Protective cloths and padding: Soft, non-abrasive cloths and padding are used to protect the vehicle's paint and surrounding panels from scratches or damage during the repair process.

d. Lubricants and rust penetrants: These substances are used to lubricate tools, reduce friction, and facilitate access to dents in tight or rust-prone areas.

By familiarizing yourself with these essential PDR tools and equipment, you'll be better prepared to tackle a wide range of dent repair scenarios. In the following section, we'll explore the step-by-step process of performing paintless dent repair, from initial assessment to final quality control.

Step-by-Step PDR Process

Paintless dent repair is a methodical process that requires skill, patience, and attention to detail. By following a step-by-step approach, technicians can ensure consistent, high-quality results and minimize the risk of damage to the vehicle's paint or panel. In this section, we'll break down the PDR process into its key stages, providing a comprehensive guide to successful dent removal.

Assessment and Planning
The first step in any PDR job is to thoroughly assess the damage and develop a clear plan of action. This initial assessment is crucial for determining the best approach, selecting the appropriate tools, and estimating the time and effort required for the repair.

Key steps in the assessment and planning stage:
a. Identify the type and severity of the dent: Use dent lights and visual inspection to determine the dent's depth, size, and location, as well as any associated paint damage or access challenges.

b. Document the damage: Use a high-resolution camera to take clear, well-lit photos of the dent from multiple angles. These photos will serve as a reference throughout the repair process and can be used to showcase the final results to the customer.

c. Select the appropriate tools and techniques: Based on the assessment, choose the most suitable PDR tools and techniques for the specific dent.

Consider factors such as the dent's location, depth, and accessibility when making your selection.

d. Develop a repair plan: Outline the step-by-step process you'll follow to remove the dent, including the order in which you'll tackle different areas and any special considerations or challenges you anticipate.

Gaining Access to the Dent
Once you've assessed the damage and developed a repair plan, the next step is to gain access to the backside of the dent. This typically involves carefully removing interior trim panels, insulation, or other components that obstruct access to the dent.

Key steps in gaining access to the dent:
a. Identify the best access point: Determine the most direct and least invasive route to the backside of the dent. This may involve removing nearby panels, drilling small access holes, or using existing openings in the vehicle's structure.

b. Remove obstructing components: Carefully remove any interior trim, insulation, or other components that block access to the dent. Use proper tools and techniques to avoid damaging these components or the surrounding area.

c. Protect the work area: Place protective cloths or padding around the access point to prevent scratches or damage to the vehicle's interior during the repair process.

Applying Pressure and Manipulating the Dent
With access to the backside of the dent established, the core of the PDR process begins. Using a combination of dent pushing tools, glue pulling systems, and gentle tapping techniques, the technician gradually coaxes the dent out, restoring the panel to its original shape.

Key steps in applying pressure and manipulating the dent:
a. Start with the deepest part of the dent: Begin by applying pressure to the deepest portion of the dent using an appropriate pushing tool. Use slow, controlled movements to gradually push the dent outward.

b. Work your way outward: As the deepest part of the dent begins to come out, move your tool toward the edges of the dent, applying steady pressure to blend the repair area with the surrounding panel.

c. Use glue pulling for challenging areas: If certain areas of the dent are difficult to access or manipulate with pushing tools, use a glue pulling system to gently lift the dent from the front side of the panel.

d. Monitor progress with dent lights: Regularly check your progress using dent lights to identify any remaining low spots, high spots, or irregularities in the repair area.

Blending and Finessing the Repair

As the dent nears its original shape, the focus shifts to blending and finessing the repair area to achieve a seamless, invisible result. This stage involves using dent taps, knockdown tools, and fine-tuned pushing techniques to level out any remaining imperfections.

Key steps in blending and finessing the repair:

a. Identify high and low spots: Use dent lights to pinpoint any remaining high or low spots in the repair area that require further attention.

b. Tap down high spots: Using a slapper file or blending hammer, gently tap down any raised areas to create a smooth, even surface.

c. Finesse low spots: For minor low spots, use fine-tipped pushing tools or glue pulling tabs to gently lift the area, blending it with the surrounding panel.

d. Check for paint damage: Inspect the repair area for any signs of paint damage, such as cracks or chips, that may have occurred during the dent removal process. If necessary, use touch-up paint or other techniques to restore the paint's integrity.

Quality Control and Reassembly

The final stage of the PDR process involves a thorough quality control check and the reassembly of any components removed during the repair. This stage ensures that the repair meets the highest standards of quality and that the vehicle is returned to its pre-damage condition.

Key steps in quality control and reassembly:

a. Inspect the repair under various lighting conditions: View the repair area under different lighting angles and intensities to ensure that the dent has been completely removed and that the panel is smooth and even.

b. Check for tool marks or other imperfections: Carefully examine the repair area for any signs of tool marks, scratches, or other imperfections that may have occurred during the repair process. If necessary, use fine-grit sandpaper or polishing compounds to remove these marks.

c. Reassemble removed components: Carefully reinstall any interior trim, insulation, or other components that were removed to access the dent. Ensure that all components are properly aligned and securely fastened.

d. Clean the work area: Remove any protective cloths, padding, or debris from the work area, leaving the vehicle's interior clean and tidy.

e. Document the completed repair: Take post-repair photos of the vehicle to showcase the quality of your work and provide a visual record of the successful dent removal.

By following this step-by-step PDR process, technicians can approach dent repair in a structured, efficient manner, ensuring optimal results and customer satisfaction. As you gain experience and refine your techniques, you'll develop a keen eye for detail and a steady hand, allowing you to tackle even the most challenging dent repairs with confidence and skill.

Advanced PDR Techniques for Challenging Dents

While the basic principles of paintless dent repair remain constant, some dents present unique challenges that require advanced techniques and specialized tools. These challenging dents may be located in hard-to-reach areas, have complex shapes, or involve unconventional materials. In this section, we'll explore advanced PDR techniques that can help you tackle these difficult repairs with confidence and skill.

Glue Pulling for Tight Access Areas

One of the most common challenges in PDR is accessing dents in tight, confined spaces where traditional pushing tools may not fit or have sufficient leverage. In these situations, glue pulling techniques can be a valuable alternative, allowing you to remove the dent from the front side of the panel.

Advanced glue pulling techniques for tight access areas:

a. Use specialized glue tabs: Select glue tabs that are specifically designed for tight access areas, such as those with narrow profiles or flexible backing materials. These tabs can be inserted into smaller gaps and conform to the contours of the panel.

b. Employ multiple tabs: For larger or more complex dents, use multiple glue tabs to distribute the pulling force evenly across the dent. This approach can help prevent further distortion of the panel and ensure a smooth, even repair.

c. Utilize mini bridge pullers: Mini bridge pullers are compact, portable tools that can be used to apply controlled pulling force to glue tabs in tight spaces. These tools allow for precise dent removal without the need for large, cumbersome equipment.

Leveraging Paintless Dent Removal
Leveraging is a technique that involves using the existing shape and contours of the panel to your advantage when removing dents. By carefully selecting your pushing or pulling points and using the panel's natural tension, you can often remove challenging dents with less effort and greater precision.

Advanced leveraging techniques for PDR:
a. Identify key leverage points: Carefully examine the dent and the surrounding panel to identify areas where the metal is naturally inclined to flex or bend. These points can serve as ideal locations for applying pushing or pulling force.

b. Use the panel's contours: Look for curves, ridges, or other structural features of the panel that can be used to your advantage when removing the dent. By pushing or pulling in the direction of these contours, you can often remove the dent with less resistance.

c. Apply controlled force: When leveraging the panel's shape, be careful to apply force gradually and controllably to avoid overworking the metal or creating new distortions. Use dent lights and visual inspection to monitor your progress and make adjustments as needed.

Metal Shrinking for Oversized Dents

In some cases, large or deep dents may cause the metal to stretch or expand, making it difficult to restore the panel to its original shape using traditional PDR techniques. In these situations, metal shrinking can be used to contract the expanded metal, allowing for a more complete and accurate repair.

Advanced metal shrinking techniques for oversized dents:

a. Apply heat: Use a heat gun or other controlled heating device to gently warm the expanded metal. This process helps to relax the metal's structure and make it more pliable for shrinking.

b. Use a shrinking hammer: A shrinking hammer is a specialized tool with a textured, waffle-like face that helps to contract the metal as it cools. Gently tap the heated metal with the shrinking hammer, working from the center of the dent outward.

c. Monitor temperature and progress: Be careful not to overheat the metal, as this can cause damage to the paint or weaken the panel's structure. Use a temperature gauge and visual inspection to monitor the metal's temperature and the progress of the shrinking process.

Addressing Unconventional Materials

As automotive technology advances, vehicles increasingly incorporate unconventional materials, such as aluminum, ultra-high-strength steel, and composites. These materials can present unique challenges for PDR technicians, as they may respond differently to traditional repair techniques.

Advanced techniques for addressing unconventional materials:
a. Use material-specific tools: Select PDR tools that are specifically designed for use with the material you're working with. For example, use plastic or nylon pushers on aluminum to avoid marring the soft metal, or employ specialized glue tabs that can adhere to textured composite surfaces.

b. Adjust your technique: Unconventional materials may require a gentler touch or different approach compared to traditional steel. For example, aluminum may require slower, more controlled pushing or pulling to avoid stretching or tearing the metal.

c. Stay informed and adaptable: As new materials and technologies emerge, stay informed about the latest PDR techniques and tools designed to address these challenges. Attend training sessions, workshops, and industry events to stay up-to-date on best practices and innovative solutions.

Combining PDR with Traditional Repair Techniques
In some cases, the most effective approach to challenging dents may involve combining PDR with traditional repair techniques, such as body filling or paint touch-up. By leveraging the strengths of each method, you can achieve optimal results and restore the vehicle to its pre-damage condition.

Advanced techniques for combining PDR with traditional repair:

a. Use PDR to minimize filler use: Begin by using PDR techniques to remove as much of the dent as possible, then use a minimal amount of body filler to smooth out any remaining imperfections. This approach helps to preserve the integrity of the panel and reduce the amount of filler required.

b. Employ PDR for hail damage repair: For vehicles with extensive hail damage, use PDR to remove the majority of the dents, then address any remaining damage or paint imperfections using traditional repair methods. This hybrid approach can save time and labor compared to a full traditional repair.

c. Integrate paint touch-up with PDR: If a dent has caused minor paint damage, such as chips or scratches, use PDR to remove the dent, then carefully apply touch-up paint to restore the paint's appearance. Be sure to properly clean and prepare the surface before applying touch-up paint to ensure optimal adhesion and color matching.

By mastering these advanced PDR techniques and learning to adapt to challenging dent scenarios, you'll be well-equipped to tackle even the most complex repairs. As you gain experience and refine your skills, you'll develop a reputation as a go-to expert for difficult dent removal, setting yourself apart in a competitive industry.

Common PDR Mistakes and How to Avoid Them

As with any skilled trade, paintless dent repair involves a learning curve and the potential for mistakes, especially when first starting out or tackling challenging repairs. By understanding common PDR mistakes and learning how to avoid them, you can minimize the risk of damaging the vehicle, ensure high-quality results, and maintain a positive reputation in the industry. In this section, we'll explore some of the most frequent PDR mistakes and provide practical tips for preventing them.

Overworking the Metal
One of the most common mistakes in PDR is overworking the metal, which occurs when a technician applies excessive pressure or repeatedly works the same area, causing the metal to become stretched, weakened, or distorted.

How to avoid overworking the metal:
a. Use proper technique: Apply gentle, controlled pressure when pushing or pulling dents, and avoid using excessive force. Use the appropriate tools for the specific dent and material you're working with.

b. Monitor progress closely: Regularly check your progress using dent lights and visual inspection to avoid pushing or pulling the dent too far. Stop working the area as soon as the dent is removed and the panel is restored to its original shape.

c. Take breaks: If you're struggling with a challenging dent, take breaks to avoid frustration and the temptation to use excessive force. Step back, reassess your approach, and return to the repair with a fresh perspective.

Misdiagnosing the Dent

Another common mistake is misdiagnosing the type, severity, or extent of the dent, which can lead to using the wrong tools or techniques, or underestimating the time and effort required for the repair.

How to avoid misdiagnosing the dent:

a. Conduct a thorough assessment: Take the time to carefully examine the dent from various angles, using dent lights and other diagnostic tools to gauge its depth, size, and shape. Consider factors such as the dent's location, the panel's material, and any access challenges.

b. Document the damage: Use a high-resolution camera to take clear, well-lit photos of the dent before starting the repair. These photos can serve as a reference throughout the repair process and help you track your progress.

c. Seek second opinions: If you're unsure about the nature of the dent or the best approach for repair, don't hesitate to seek a second opinion from a more experienced technician or colleague. Collaborating with others can help you avoid costly mistakes and improve your diagnostic skills.

Neglecting Access and Clearance
Failing to properly assess and plan for access and clearance issues is another common mistake that can lead to wasted time, inefficient repairs, or damage to the vehicle's interior components.

How to avoid neglecting access and clearance:
a. Plan your approach: Before starting the repair, carefully consider the best access point for reaching the backside of the dent. Look for existing openings, removable panels, or other access routes that minimize the need for disassembly.

b. Use caution when removing components: If you must remove interior trim, insulation, or other components to access the dent, use the proper tools and techniques to avoid damaging these components or the surrounding area. Keep track of any fasteners or small parts removed during disassembly.

c. Protect the work area: Place protective cloths or padding around the access point to prevent scratches or damage to the vehicle's interior during the repair process. Be mindful of any wiring, sensors, or other delicate components that could be damaged by tools or debris.

Ignoring Paint Condition
Neglecting to assess and address any pre-existing paint damage or failing to monitor the paint's condition during the repair process can lead to unsatisfactory results or additional damage.

How to avoid ignoring paint condition:
a. Inspect the paint before starting: Before beginning the repair, carefully examine the dent and surrounding area for any pre-existing paint damage, such as chips, scratches, or cracks. Document these imperfections and discuss them with the customer before proceeding.

b. Monitor the paint during the repair: As you work the dent, regularly inspect the paint for any signs of cracking, flaking, or other damage that may occur as a result of the PDR process. If you notice any paint damage, stop the repair and reassess your approach.

c. Use paint-safe techniques: When using glue pulling or other techniques that involve adhesives or abrasives, be sure to use products and methods that are safe for the vehicle's paint. Avoid using excessive heat or pressure, which can cause the paint to blister or peel.

Rushing the Repair Process
Finally, rushing the repair process or attempting to take shortcuts can lead to subpar results, missed details, or additional damage to the vehicle.

How to avoid rushing the repair process:
a. Allow sufficient time: When scheduling PDR jobs, be realistic about the time required for each repair, taking into account factors such as the dent's size, location, and complexity. Avoid overbooking or making promises you can't keep.

b. Work methodically: Follow a systematic, step-by-step approach to each repair, taking the time to properly assess the damage, plan your approach, and execute each stage of the process with care and precision.

c. Double-check your work: Before reassembling components or returning the vehicle to the customer, take a final pass over the repair area to check for any missed details, imperfections, or remaining damage. Use dent lights and a critical eye to ensure the repair meets your high standards.

By familiarizing yourself with these common PDR mistakes and implementing strategies to avoid them, you'll be better equipped to handle a wide range of dent repair scenarios with skill, efficiency, and professionalism. Remember, the key to success in PDR is a combination of proper training, practical experience, and a commitment to continuous learning and improvement.

Chapter 3
Traditional Dent Repair Methods

Filling and Sanding Techniques

In traditional dent repair, filling and sanding techniques are essential for restoring a damaged panel to its original shape and preparing it for painting. These methods involve applying body filler to the dent, then sanding it smooth to create an even surface that blends seamlessly with the surrounding area. In this section, we'll explore the key steps and techniques involved in filling and sanding, as well as best practices for achieving professional-quality results.

Preparing the Dent for Filling
Before applying body filler, it's crucial to properly prepare the dent and surrounding area to ensure optimal adhesion and a smooth, long-lasting repair.

Steps for preparing the dent:
a. Remove loose debris: Use a wire brush or sandpaper to remove any loose paint, rust, or other debris from the dent and surrounding area. This will help the filler adhere properly and prevent future corrosion.

b. Rough sand the area: Using 80-grit sandpaper, rough sand the dent and a few inches around it to create a slightly abraded surface that will help the filler bond to the metal. Be careful not to sand too aggressively, as this can remove too much of the original paint or create deep scratches.

c. Clean the area: Use a solvent-based cleaner or degreaser to remove any dirt, oil, or residue from the sanded area. This will help ensure a clean, contaminant-free surface for the filler to adhere to.

d. Apply a metal conditioner (optional): If the dent is on a bare metal surface or has exposed metal due to sanding, apply a metal conditioner or primer to help prevent rust and promote better filler adhesion.

Mixing and Applying Body Filler
Body filler, also known as bondo or putty, is a two-part polyester resin that is mixed with a hardener to create a thick, moldable paste that can be used to fill and shape dents.

Steps for mixing and applying body filler:
a. Select the appropriate filler: Choose a high-quality body filler that is suitable for the size and depth of the dent, as well as the material of the panel. Some fillers are designed for shallow dents, while others are better suited for deeper damage or specific materials like aluminum.

b. Mix the filler and hardener: Follow the manufacturer's instructions for mixing the filler and hardener in the proper ratio. Typically, a small amount of hardener is added to a golf ball-sized portion of filler and mixed thoroughly until a uniform color is achieved.

c. Apply the filler: Using a plastic spreader or applicator, apply the mixed filler to the dent, slightly overfilling it to allow for sanding. Work quickly, as the filler will begin to harden within a few minutes of mixing.

d. Shape the filler: Before the filler fully hardens, use the spreader to roughly shape it to match the contours of the surrounding panel. This will save time and effort during the sanding process.

Sanding the Body Filler
Once the body filler has fully hardened, it must be sanded smooth to create an even surface that blends with the surrounding panel.

Steps for sanding the body filler:
a. Rough sand the filler: Using 80-grit sandpaper, remove any excess filler and roughly shape the repair area to match the contours of the panel. Sand in a crisscross pattern to avoid creating deep scratches or grooves.

b. Gradually decrease sandpaper grit: Progress to 120-grit, then 220-grit sandpaper to further smooth and refine the shape of the filler. Use a sanding block or interface pad to maintain an even surface and avoid oversanding in any one area.

c. Feather-edge the repair: Using 320-grit sandpaper, gently sand the edges of the repair area to create a smooth, gradual transition between the filler and the surrounding paint. This process, known as feather-edging, helps to ensure an invisible repair once the area is painted.

d. Check for imperfections: Inspect the sanded area for any pinholes, low spots, or other imperfections. If necessary, apply a thin layer of glazing putty or spot putty to fill any minor defects, then sand smooth with 400-grit sandpaper.

Best Practices for Filling and Sanding

To achieve the best possible results when filling and sanding dents, keep these best practices in mind:

a. Work in a well-ventilated area: Body filler and sanding dust can be harmful if inhaled, so always work in a well-ventilated area and wear a respirator or dust mask.

b. Use a guide coat: Between sanding steps, apply a thin layer of contrasting color spray paint or primer to the repair area. This guide coat will help you identify high and low spots as you sand, ensuring a more even and accurate repair.

c. Keep the area clean: Regularly remove sanding dust and debris from the repair area with compressed air or a tack cloth to avoid contaminating the filler or creating an uneven surface.

d. Avoid oversanding: Be careful not to sand too aggressively or remove too much of the filler, as this can create low spots or expose the original dent. Take your time and sand gradually, checking your progress frequently.

e. Practice on scrap panels: Before tackling a real dent repair, practice your filling and sanding techniques on scrap metal panels or old body parts. This will help you develop a feel for the materials and tools, as well as refine your skills.

By mastering these filling and sanding techniques and following best practices, you'll be able to achieve seamless, professional-quality dent repairs that are ready for priming and painting. In the next section, we'll explore the process of applying primer and paint to complete the repair and restore the panel to its original appearance.

Using Body Filler for Dent Repair

Body filler, also known as bondo or putty, is a versatile and essential material for repairing dents that are too deep or extensive to be addressed with paintless dent repair (PDR) techniques alone. When used correctly, body filler can help restore a damaged panel to its original shape and prepare it for painting. In this section, we'll delve into the process of using body filler for dent repair, including surface preparation, application techniques, and tips for achieving the best results.

Choosing the Right Body Filler
Before beginning the repair process, it's important to select the appropriate type of body filler for the specific dent and material you're working with.

Factors to consider when choosing body filler:
a. Depth of the dent: For shallow dents, a lightweight, easy-sanding filler is often sufficient. Deeper dents may require a heavier, more durable filler that can be built up in layers.

b. Material of the panel: Some body fillers are specifically formulated for use on certain materials, such as aluminum or galvanized steel. Be sure to choose a filler that is compatible with the substrate to ensure proper adhesion and durability.

c. Desired finish: If the repair area will be visible on a high-end vehicle or a show car, choose a premium-quality filler that can be sanded to an ultra-smooth finish. For less critical repairs, a standard-grade filler may suffice.

Preparing the Surface for Body Filler
Proper surface preparation is crucial for ensuring that the body filler adheres properly and creates a long-lasting repair.

Steps for preparing the surface:
a. Remove any loose debris: Use a wire brush, sandpaper, or abrasive disc to remove any loose paint, rust, or other debris from the dent and surrounding area. This will help create a clean, stable surface for the filler to bond to.

b. Rough sand the area: Using 40- or 80-grit sandpaper, rough up the dent and a few inches around it to create a slightly abraded surface that will help the filler adhere. Be careful not to sand too aggressively, as this can thin the metal and weaken the panel.

c. Clean the surface: Use a solvent-based cleaner or degreaser to remove any dirt, oil, or residue from the sanded area. This will help ensure a contaminant-free surface that promotes optimal filler adhesion.

d. Apply a metal conditioner (optional): If the dent is on a bare metal surface or has exposed metal due to sanding, apply a metal conditioner or primer to help prevent rust and promote better filler adhesion.

Mixing and Applying the Body Filler
Once the surface is properly prepared, it's time to mix and apply the body filler.

Steps for mixing and applying body filler:
a. Read the manufacturer's instructions: Each brand of body filler may have slightly different mixing ratios and application techniques. Be sure to read and follow the instructions carefully to ensure the best results.

b. Mix the filler and hardener: Dispense the appropriate amount of filler onto a clean, non-porous surface. Add the hardener according to the manufacturer's recommended ratio (usually a small ribbon per golf ball-sized amount of filler). Mix the filler and hardener thoroughly using a plastic spreader until a uniform color is achieved.

c. Apply the filler to the dent: Using a plastic spreader or applicator, apply the mixed filler to the dent, slightly overfilling it to allow for sanding. Work quickly and smoothly, as the filler will begin to harden within a few minutes of mixing.

d. Shape the filler: Before the filler fully hardens, use the spreader to roughly shape it to match the contours of the surrounding panel. This will save time and effort during the sanding process.

Sanding and Finishing the Repair
After the body filler has fully cured (typically 15-30 minutes, depending on the product), it must be sanded and finished to create a smooth, invisible repair.

Steps for sanding and finishing the repair:
a. Rough sand the filler: Using 80-grit sandpaper, remove any excess filler and roughly shape the repair area to match the contours of the panel. Sand in a crisscross pattern to avoid creating deep scratches or grooves.

b. Gradually decrease sandpaper grit: Progress to 120-grit, then 220-grit sandpaper to further smooth and refine the shape of the filler. Use a sanding block or interface pad to maintain an even surface and avoid oversanding in any one area.

c. Feather-edge the repair: Using 320-grit sandpaper, gently sand the edges of the repair area to create a smooth, gradual transition between the filler and the surrounding paint. This process, known as feather-edging, helps to ensure an invisible repair once the area is painted.

d. Check for imperfections: Inspect the sanded area for any pinholes, low spots, or other imperfections. If necessary, apply a thin layer of glazing putty or spot putty to fill any minor defects, then sand smooth with 400-grit sandpaper.

e. Prime and paint the repair: Once the filler is smoothly sanded and free of imperfections, apply a high-quality primer to the repair area, then sand it smooth with 400-grit sandpaper. Follow up with the appropriate basecoat and clearcoat paint to complete the repair and blend it with the surrounding panel.

Tips for Using Body Filler Successfully

To achieve the best results when using body filler for dent repair, keep these tips in mind:

a. Use high-quality materials: Invest in premium-grade body filler, hardener, and sandpaper to ensure optimal performance, sanding efficiency, and durability of the repair.

b. Apply filler in layers: For deep dents, apply the filler in several thin layers, allowing each layer to cure and sanding it smooth before applying the next. This will help prevent excessive shrinkage and cracking of the filler.

c. Avoid overworking the filler: Mix only as much filler as you can apply within the product's working time (usually 3-5 minutes). Overworking or attempting to apply partially cured filler can result in poor adhesion and an inferior repair.

d. Use a guide coat: Between sanding steps, apply a thin layer of contrasting color spray paint or primer to the repair area. This guide coat will help you identify high and low spots as you sand, ensuring a more even and accurate repair.

e. Practice proper safety measures: Body filler dust and vapors can be harmful if inhaled, so always work in a well-ventilated area and wear a respirator or dust mask. Wear gloves to protect your hands from the chemicals in the filler and hardener.

By understanding the properties of body filler, following proper application techniques, and adhering to best practices, you can achieve professional-quality dent repairs that restore damaged panels to their original shape and appearance. In the next section, we'll explore the process of priming and painting the repaired area to complete the job.

Priming and Painting Repaired Areas

After filling and sanding a dent, the repaired area must be primed and painted to protect the underlying metal, ensure a smooth and durable finish, and blend the repair seamlessly with the surrounding panel. Proper priming and painting techniques are essential for achieving a professional-quality repair that looks factory-original. In this section, we'll explore the steps involved in priming and painting repaired areas, as well as best practices for achieving optimal results.

Choosing the Right Primer
Primer is a crucial component of the painting process, as it provides a stable foundation for the topcoat and helps to prevent corrosion and adhesion issues.

Factors to consider when choosing a primer:
a. Compatibility with the substrate: Select a primer that is compatible with the material you're working with, whether it's bare metal, body filler, or existing paint. Some primers are specifically formulated for certain substrates, such as etching primers for bare metal or high-build primers for body filler.

b. Color and opacity: Choose a primer color that is appropriate for the topcoat you'll be applying. Light gray or white primers are suitable for most light-colored topcoats, while darker gray or black primers are better for darker colors. Consider the opacity of the primer as well, as some high-opacity primers can help to conceal minor imperfections and color variations.

c. Environmental considerations: Opt for a low-VOC (volatile organic compound) primer if you're working in an area with strict environmental regulations or if you prefer a more eco-friendly approach. Water-based primers are also available as an alternative to solvent-based products.

Preparing the Surface for Primer
Before applying primer, the repaired area must be thoroughly cleaned and prepared to ensure optimal adhesion and a smooth finish.

Steps for preparing the surface:
a. Clean the area: Use a solvent-based cleaner or degreaser to remove any dirt, oil, or residue from the repaired area and surrounding panel. Pay special attention to the edges of the repair, where sanding dust and debris can accumulate.

b. Scuff the surrounding paint: Using a fine-grit sandpaper (400-600 grit) or a scuff pad, gently abrade the paint surrounding the repair area to create a slight texture that will help the primer adhere. Be careful not to sand through the existing paint or create deep scratches.

c. Mask off the surrounding area: Apply masking tape and paper or plastic sheeting to protect the surrounding panel from overspray. Make sure the edges of the tape are pressed down firmly to prevent primer or paint from seeping underneath.

d. Apply a pre-painting prep: If desired, apply a pre-painting preparation product, such as a wax and grease remover or a plastic adhesion promoter, to the repaired area. These products can help to further clean the surface and improve the adhesion of the primer.

Applying the Primer
Once the surface is properly prepared, the primer can be applied using a spray gun or aerosol can.

Steps for applying primer:
a. Mix and strain the primer: If using a sprayable primer, mix it according to the manufacturer's instructions and strain it through a fine-mesh filter to remove any lumps or debris. If using an aerosol primer, shake the can thoroughly before use.

b. Set up the spray equipment: If using a spray gun, adjust the air pressure, fluid flow, and fan pattern according to the manufacturer's recommendations for the specific primer you're using. If using an aerosol can, hold it 8-10 inches away from the surface and keep it moving to avoid heavy application in any one area.

c. Apply the primer: Begin by applying a light, even coat of primer to the repaired area and a few inches beyond. Allow this first coat to dry according to the manufacturer's instructions, then apply a second coat if necessary to achieve full coverage and a smooth finish.

d. Sand the primer: Once the primer has fully dried (usually 30-60 minutes, depending on the product), sand it lightly with 400-600 grit sandpaper to remove any imperfections and create a smooth surface for the topcoat. Use a tack cloth to remove any sanding dust before proceeding.

Applying the Topcoat
With the primer applied and sanded, the repaired area is ready for the topcoat, which typically consists of a basecoat color and a clearcoat for added depth and protection.

Steps for applying the topcoat:
a. Mix and strain the paint: Mix the basecoat color according to the manufacturer's instructions, using a paint mixing system or formula to match the existing color of the vehicle. Strain the mixed paint through a fine-mesh filter to remove any lumps or debris.

b. Set up the spray equipment: Adjust the spray gun settings according to the manufacturer's recommendations for the specific paint you're using. Make sure the gun is clean and free of any residue from previous painting sessions.

c. Apply the basecoat: Begin by applying a light, even coat of basecoat color to the primed area, extending slightly beyond the repair to ensure a smooth blend with the surrounding paint. Allow this first coat to dry according to the manufacturer's instructions, then apply additional coats as necessary to achieve full coverage and color match.

d. Apply the clearcoat: Once the basecoat has dried (usually 15-30 minutes, depending on the product), apply 2-3 coats of clearcoat over the entire repaired area, allowing each coat to dry before applying the next. Make sure to overlap the edges of the repair slightly to ensure a smooth transition.

e. Allow the paint to cure: After applying the final coat of clearcoat, allow the paint to cure according to the manufacturer's instructions. This may take anywhere from a few hours to several days, depending on the product and environmental conditions.

Best Practices for Priming and Painting
To achieve the best possible results when priming and painting repaired areas, keep these tips in mind:

a. Use high-quality materials: Invest in premium-grade primers, paints, and clearcoats to ensure optimal durability, color match, and finish quality. Avoid using cheap or generic products, as they may not perform as well or provide long-lasting results.

b. Maintain a clean work environment: Make sure your workspace is clean, well-ventilated, and free of dust, dirt, and other contaminants that can compromise the quality of the repair. Use a tack cloth to remove any debris from the surface before priming or painting.

c. Follow the manufacturer's instructions: Always read and follow the manufacturer's instructions for mixing, application, and drying times for the specific products you're using. Don't try to rush the process or take shortcuts, as this can lead to poor adhesion, uneven coverage, or other issues.

d. Practice proper safety measures: Wear appropriate personal protective equipment (PPE), such as a respirator, gloves, and eye protection, when working with primers, paints, and solvents. Make sure your workspace is well-ventilated and avoid smoking or using open flames near painting materials.

e. Allow adequate drying and curing time: Don't rush the drying and curing process for primers and paints. Allow each coat to dry thoroughly before applying the next, and give the final finish plenty of time to cure before subjecting it to any stress or exposure to the elements.

By following these priming and painting techniques and adhering to best practices, you can achieve professional-quality results that blend seamlessly with the surrounding panel and provide long-lasting protection for the repaired area. With practice and attention to detail, your dent repairs will be virtually invisible, restoring the vehicle's appearance and value.

Blending New Paint with Existing Paint

When repairing a dent or damage on a vehicle, one of the most challenging aspects is achieving a seamless blend between the newly painted area and the existing paint. Blending is the process of gradually fading the new paint into the surrounding area to create a smooth, invisible transition. Proper blending techniques are essential for ensuring a professional-quality repair that matches the color, texture, and sheen of the original paint. In this section, we'll explore the key steps and techniques involved in blending new paint with existing paint.

Understanding Color Matching
Before attempting to blend new paint, it's crucial to ensure that the color of the new paint matches the existing paint as closely as possible.

Factors to consider when color matching:
a. Paint code: Every vehicle has a specific paint code that identifies the exact color and formulation of the original paint. This code can usually be found on a sticker or label in the engine compartment, door jamb, or glove box. Use this code to order the correct paint from a supplier.

b. Color variations: Even with the correct paint code, there may be slight variations in color due to factors such as age, sun exposure, and previous repairs. To account for these variations, it's essential to compare the new paint to the existing paint in different lighting conditions and angles.

c. Blendable vs. non-blendable colors: Some colors, such as solid whites, blacks, and reds, are easier to blend than others, such as metallics, pearls, and tricoats. Be aware of the type of color you're working with and adjust your blending technique accordingly.

Preparing the Area for Blending
Before applying the new paint, the surrounding area must be properly prepared to ensure a smooth and seamless blend.

Steps for preparing the area:
a. Clean the surface: Use a solvent-based cleaner or degreaser to remove any dirt, oil, or residue from the repair area and the surrounding panel. This will help the new paint adhere properly and prevent contamination.

b. Scuff the surrounding paint: Using a fine-grit sandpaper (1000-1500 grit) or a scuff pad, gently abrade the paint in the blending area to create a slight texture that will help the new paint adhere and blend smoothly. Be careful not to sand through the clearcoat or create deep scratches.

c. Mask off the adjacent panels: Apply masking tape and paper or plastic sheeting to protect the adjacent panels from overspray. Make sure the edges of the tape are pressed down firmly to prevent paint from seeping underneath.

d. Apply a blending solvent: If desired, apply a blending solvent or reducer to the edges of the repair area to help the new paint flow and blend more smoothly with the existing paint. Be careful not to apply too much solvent, as this can cause the paint to run or sag.

Applying the New Paint
Once the area is prepared, the new paint can be applied using a spray gun or aerosol can.

Steps for applying the new paint:
a. Mix and strain the paint: Mix the paint according to the manufacturer's instructions, using a paint mixing system or formula to match the existing color. Strain the mixed paint through a fine-mesh filter to remove any lumps or debris.

b. Set up the spray equipment: Adjust the spray gun settings according to the manufacturer's recommendations for the specific paint you're using. For blending, you may need to use a lower air pressure and a wider fan pattern to achieve a softer, more gradual transition.

c. Apply the basecoat: Begin by applying the new basecoat color to the repair area, extending slightly beyond the scuffed area to ensure a smooth blend. Apply the paint in thin, even coats, allowing each coat to dry before applying the next. As you approach the edges of the repair, gradually lighten your pressure on the spray gun trigger to create a soft, feathered edge.

d. Blend the new paint: To blend the new paint with the existing paint, use a blending technique such as the "mist coat" or the "wet bed" method. For the mist coat method, apply a very light, misty coat of the new color over the entire blending area, allowing it to dry before applying the clearcoat. For the wet bed method, apply a coat of blending solvent or reducer to the edges of the repair area, then immediately apply the new color, allowing it to flow and blend with the existing paint.

e. Apply the clearcoat: Once the basecoat is blended and dried, apply 2-3 coats of clearcoat over the entire repair area, extending slightly beyond the blended area to ensure a smooth transition. Allow each coat to dry before applying the next, and make sure to follow the manufacturer's instructions for flash times and drying times.

Tips for Achieving a Seamless Blend
To achieve the best possible results when blending new paint with existing paint, keep these tips in mind:

a. Practice on scrap panels: Before attempting to blend paint on a customer's vehicle, practice your technique on scrap panels or old hoods. This will help you develop a feel for the paint and the blending process, and allow you to make mistakes without consequence.

b. Use high-quality materials: Invest in premium-grade paints, clearcoats, and blending solvents to ensure optimal color match, durability, and blendability. Avoid using cheap or generic products, as they may not perform as well or provide long-lasting results.

c. Work in a well-lit area: Make sure your workspace is brightly lit and free of shadows or glare. This will help you see the color and texture of the paint more clearly and identify any areas that need additional blending.

d. Take your time: Blending paint is a skill that requires patience and attention to detail. Don't rush the process or try to cut corners, as this can lead to a poor blend or visible transition lines. Take your time and focus on achieving a smooth, gradual transition between the new and existing paint.

e. Check your work from different angles: After blending the paint, check your work from different angles and in different lighting conditions to ensure a seamless blend. If necessary, apply additional mist coats or blending solvents to refine the transition and achieve a perfect match.

By mastering these blending techniques and following best practices, you can achieve professional-quality paint repairs that are virtually invisible and match the original finish of the vehicle. With practice and experience, you'll develop a keen eye for color matching and a steady hand for blending, allowing you to tackle even the most challenging paint repairs with confidence.

Chapter 4
Bumper Repair and Replacement

Assessing Bumper Damage

When a vehicle's bumper is damaged, whether due to a collision, parking mishap, or other incident, it's essential to thoroughly assess the extent of the damage to determine the most appropriate repair or replacement approach. Properly evaluating bumper damage can help you make informed decisions about the cost, time, and effort required to restore the bumper to its pre-accident condition. In this section, we'll explore the key factors to consider when assessing bumper damage and provide guidance on how to determine the best course of action.

Identifying the Type of Bumper Material
The first step in assessing bumper damage is to identify the type of material the bumper is made from, as this will influence the repair or replacement options available.

Common bumper materials include:
a. Plastic: Most modern vehicles feature bumpers made from thermoplastic materials, such as polypropylene or polyurethane. These bumpers are lightweight, flexible, and resistant to minor impacts, but they can crack, scratch, or dent under more severe stress.

b. Metal: Some older vehicles or specialty cars may have bumpers made from chrome-plated steel or aluminum. These bumpers are more rigid and durable than plastic, but they can be more challenging to repair if bent or dented.

c. Fiberglass: Less common than plastic or metal, fiberglass bumpers are sometimes found on custom or classic vehicles. These bumpers are lightweight and can be molded into unique shapes, but they are brittle and prone to cracking or shattering on impact.

Evaluating the Extent of the Damage
Once you've identified the bumper material, the next step is to carefully inspect the bumper and assess the extent of the damage.

Key factors to consider when evaluating bumper damage:
a. Location and size of the damage: Note the location and size of any cracks, dents, or scratches on the bumper. Damage that is confined to a small area may be easier to repair than more extensive damage that spans multiple sections of the bumper.

b. Depth of the damage: Determine how deep any cracks, dents, or gouges extend into the bumper material. Shallow damage that only affects the surface may be suitable for repair, while deeper damage that penetrates through the bumper or compromises its structural integrity may require replacement.

c. Condition of the paint: Evaluate the condition of the paint on the bumper, noting any chips, scrapes, or peeling. Minor paint damage may be addressed with touch-up or repainting, while more severe damage may necessitate sanding and refinishing the entire bumper.

d. Presence of internal damage: Check for any signs of damage to the bumper's internal structure, such as broken mounting points, cracked reinforcement bars, or damaged foam absorbers. Internal damage may not be visible from the exterior but can compromise the bumper's ability to absorb impact and protect the vehicle.

Considering Repair vs. Replacement

Based on your assessment of the bumper damage, you'll need to determine whether repair or replacement is the most appropriate course of action.

Factors to consider when deciding between repair and replacement:

a. Extent of the damage: If the damage is minor and confined to a small area, such as a shallow scratch or small dent, repair may be the most cost-effective and efficient option. However, if the damage is extensive, deep, or spans multiple areas of the bumper, replacement may be necessary to restore the bumper's appearance and structural integrity.

b. Material of the bumper: The type of bumper material will influence the feasibility and cost of repair. Plastic bumpers are generally easier and less expensive to repair than metal or fiberglass, as they can often be filled, sanded, and repainted without the need for extensive bodywork. Metal and fiberglass bumpers may require more specialized repair techniques or may be more cost-effective to replace if severely damaged.

c. Age and value of the vehicle: For older or lower-value vehicles, the cost of replacing a severely damaged bumper may exceed the car's worth. In these cases, repair may be the more economical option, even if the results are not perfect. For newer or higher-value vehicles, however, investing in a proper repair or replacement may be justified to maintain the car's appearance and resale value.

d. Availability of replacement parts: The availability and cost of replacement bumpers will vary depending on the make, model, and year of the vehicle. For some cars, aftermarket or used bumpers may be readily available at a lower cost than OEM parts, making replacement a more attractive option. For others, replacement bumpers may be rare or expensive, making repair a more practical choice.

Estimating Repair or Replacement Costs
Once you've determined whether repair or replacement is the best approach, the next step is to estimate the costs involved to help the customer make an informed decision.

Factors to consider when estimating bumper repair or replacement costs:
a. Labor costs: Estimate the time required to complete the repair or replacement, taking into account any necessary disassembly, preparation, painting, and reassembly. Multiply the estimated labor hours by your shop's hourly rate to determine the labor cost.

b. Material costs: Calculate the cost of any materials needed for the repair or replacement, such as body filler, sandpaper, paint, or a new bumper. Be sure to include any additional components that may need to be replaced, such as mounting brackets, absorbers, or trim pieces.

c. Sublet costs: If any portion of the repair or replacement needs to be outsourced to a specialist, such as a paintless dent repair technician or a chrome plating shop, factor in the cost of these sublet services.

d. Overhead and markup: Don't forget to account for your shop's overhead expenses and desired profit margin when calculating the total cost of the repair or replacement. This will help ensure that the job is priced fairly and profitably.

By thoroughly assessing bumper damage and considering all relevant factors, you can provide customers with accurate recommendations and estimates for repair or replacement. In the following sections, we'll delve into the specific techniques and processes involved in repairing and replacing various types of bumpers.

Repairing Plastic Bumpers

Plastic bumpers are the most common type found on modern vehicles, thanks to their lightweight, flexible, and impact-resistant properties. When a plastic bumper sustains damage, such as cracks, dents, or scratches, it can often be repaired using specialized techniques and materials. In this section, we'll explore the key steps and considerations involved in repairing plastic bumpers, from preparation to finishing.

Cleaning and Preparing the Damaged Area
Before beginning any repair work, it's essential to thoroughly clean and prepare the damaged area to ensure proper adhesion and a seamless finish.

Steps for cleaning and preparing the damaged area:
a. Wash the bumper: Use a mild detergent and water to wash the entire bumper, removing any dirt, grime, or debris. Pay special attention to the damaged area, ensuring that it is free of any contaminants that could interfere with the repair process.

b. Dry the bumper: Use a clean microfiber towel or compressed air to dry the bumper completely. Avoid using paper towels or rags that could leave lint or fibers behind.

c. Remove loose or damaged material: Using a plastic scraper or utility knife, carefully remove any loose paint, plastic, or other debris from the damaged area. Be careful not to cause further damage or enlarge the affected area.

d. Sand the damaged area: Using 80-grit sandpaper, gently sand the edges of the damaged area to create a slightly roughened surface that will help the repair material adhere. Be sure to sand a few inches beyond the visible damage to ensure a smooth transition.

e. Clean the sanded area: Use a plastic cleaner or degreaser to remove any sanding dust or residue from the damaged area. This will help ensure a clean, contaminant-free surface for the repair.

Applying Plastic Repair Material
Once the damaged area is properly cleaned and prepared, the next step is to fill and reshape the area using a specialized plastic repair material.

Steps for applying plastic repair material:
a. Select the appropriate repair material: Choose a plastic repair material that is compatible with the type of plastic used in the bumper. Some common options include two-part epoxy fillers, thermoplastic repair kits, and plastic welding systems.

b. Mix the repair material: If using a two-part filler or repair kit, mix the components according to the manufacturer's instructions, ensuring a thorough and even blend. If using a plastic welding system, load the appropriate plastic welding rod into the welding gun.

c. Apply the repair material: Using a plastic spreader or the provided applicator, apply the mixed repair material to the damaged area, slightly overfilling it to allow for sanding and shaping. If using a plastic welding system, carefully melt and apply the welding rod to the damaged area, building up the material in layers until the area is slightly overfilled.

d. Allow the repair material to cure: Let the repair material cure according to the manufacturer's instructions. This may take anywhere from a few minutes to several hours, depending on the type of material used and the ambient temperature and humidity.

Shaping and Sanding the Repaired Area
After the repair material has fully cured, it must be shaped and sanded to match the original contours of the bumper and create a smooth, seamless transition with the surrounding area.

Steps for shaping and sanding the repaired area:
a. Rough sand the repair material: Using 80-grit sandpaper, sand down the excess repair material until it is level with the surrounding bumper. Use a sanding block or a flexible sanding pad to ensure an even surface and avoid creating low spots or gouges.

b. Shape the repair area: Using a finer grit sandpaper (120-grit to 220-grit), continue sanding the repair area, shaping it to match the original contours and curves of the bumper. Check your progress frequently to avoid oversanding or altering the bumper's shape.

c. Feather-edge the repair: Using 320-grit or finer sandpaper, gently sand the edges of the repair area to create a smooth, gradual transition with the surrounding bumper. This process, known as feather-edging, helps to ensure an invisible repair once the area is painted.

d. Check for imperfections: Inspect the sanded area for any pinholes, low spots, or other imperfections. If necessary, apply a thin layer of glazing putty or spot filler to address any minor flaws, then sand smooth with 400-grit or finer sandpaper.

Priming and Painting the Repaired Bumper
With the repair area properly shaped and sanded, the final step is to prime and paint the bumper to restore its original appearance.

Steps for priming and painting the repaired bumper:
a. Clean the sanded area: Use a tack cloth or compressed air to remove any sanding dust or debris from the repaired area and surrounding bumper. Ensure that the surface is clean and dry before proceeding.

b. Apply plastic adhesion promoter: If recommended by the primer or paint manufacturer, apply a plastic adhesion promoter to the repaired area to enhance the bond between the plastic and the primer. Allow the adhesion promoter to dry according to the manufacturer's instructions.

c. Apply primer: Using a spray gun or aerosol primer, apply a high-quality, compatible primer to the repaired area and a few inches beyond, ensuring even coverage and a smooth finish. Allow the primer to dry and cure according to the manufacturer's instructions.

d. Sand the primed area: Using 600-grit or finer sandpaper, lightly sand the primed area to remove any imperfections and create a smooth surface for the topcoat. Use a tack cloth to remove any sanding dust.

e. Apply the topcoat: Using a spray gun or aerosol paint matched to the bumper's original color, apply the topcoat in even, overlapping passes, extending slightly beyond the repaired area to ensure a seamless blend. Apply multiple coats as needed, allowing each coat to dry before applying the next.

f. Clear coat and buff (optional): If desired, apply a compatible clear coat over the repaired area to enhance the depth and gloss of the paint. Once the clear coat has cured, use a fine-grit polishing compound and a buffing pad to remove any minor imperfections and achieve a smooth, glossy finish.

Tips for Successful Plastic Bumper Repair
To achieve the best possible results when repairing plastic bumpers, keep these tips in mind:

a. Work in a well-ventilated area: Plastic repair materials, primers, and paints can emit strong fumes, so always work in a well-ventilated area and wear a respirator or mask when necessary.

b. Use compatible products: Ensure that all repair materials, primers, and paints used are compatible with the specific type of plastic used in the bumper. Consult the manufacturer's guidelines or a compatibility chart to avoid adverse reactions or poor adhesion.

c. Take your time: Rushing the repair process can lead to mistakes or subpar results. Allow adequate time for each step, from cleaning and preparation to curing and sanding, to ensure a high-quality, long-lasting repair.

d. Practice on scrap bumpers: If you're new to plastic bumper repair, practice your techniques on scrap or salvage bumpers before attempting a repair on a customer's vehicle. This will help you develop your skills and gain confidence in your abilities.

By following these steps and tips, you can successfully repair most types of damage to plastic bumpers, restoring their appearance and structural integrity. With practice and attention to detail, your plastic bumper repairs will be virtually undetectable, ensuring customer satisfaction and a professional reputation.

Fixing Cracked or Broken Bumpers

While some bumper damage, such as minor dents or scratches, can be repaired using the techniques described in the previous section, more severe damage, such as cracks or breaks, may require additional steps or even partial replacement. In this section, we'll explore the specific challenges and techniques involved in fixing cracked or broken bumpers, with a focus on plastic bumpers, which are the most common type found on modern vehicles.

Assessing the Severity of the Damage
The first step in fixing a cracked or broken bumper is to carefully assess the severity and extent of the damage to determine the most appropriate repair approach.

Key factors to consider when assessing cracked or broken bumpers:
a. Location and size of the cracks or breaks: Note the location and size of any cracks or breaks in the bumper. Damage that is confined to a small area or a single crack may be easier to repair than more extensive damage that spans multiple areas or involves multiple cracks or breaks.

b. Depth and direction of the cracks or breaks: Evaluate the depth and direction of the cracks or breaks. Shallow, hairline cracks that don't penetrate through the bumper material may be suitable for repair, while deeper, wider cracks or breaks that go through the bumper may require partial replacement.

c. Condition of the surrounding material: Assess the condition of the plastic surrounding the cracks or breaks. If the material is brittle, crumbly, or significantly degraded, it may not be suitable for repair and may require replacement.

d. Presence of missing pieces: Check for any missing pieces or fragments from the cracked or broken area. If significant portions of the bumper are missing, repair may not be feasible, and replacement may be necessary.

Cleaning and Preparing the Damaged Area
As with any bumper repair, proper cleaning and preparation of the damaged area are essential for achieving a strong, long-lasting fix.

Steps for cleaning and preparing cracked or broken bumpers:
a. Wash and dry the bumper: Use a mild detergent and water to clean the entire bumper, then dry it thoroughly with a clean microfiber towel or compressed air.

b. Remove any loose or damaged material: Using a plastic scraper or utility knife, carefully remove any loose paint, plastic, or debris from the cracked or broken area. Be sure to remove any hanging or protruding pieces that could interfere with the repair.

c. Sand the damaged area: Using 80-grit sandpaper, gently sand the edges of the cracks or breaks to create a slightly roughened surface that will help the repair material adhere. Sand a few inches beyond the visible damage to ensure a smooth transition.

d. Clean the sanded area: Use a plastic cleaner or degreaser to remove any sanding dust, dirt, or contaminants from the damaged area and surrounding bumper.

Reinforcing and Filling the Cracks or Breaks
To restore the structural integrity of a cracked or broken bumper, it's essential to reinforce and fill the damaged area before proceeding with the standard filling and finishing steps.

Steps for reinforcing and filling cracks or breaks:
a. Install reinforcement mesh (for larger cracks or breaks): If the cracks or breaks are wide or extensive, it may be necessary to install a reinforcement mesh to provide additional strength and stability. Cut a piece of plastic reinforcement mesh slightly larger than the damaged area, then secure it to the backside of the bumper using a compatible plastic adhesive or welding system.

b. Inject plastic welding adhesive (for smaller cracks): For smaller, narrower cracks, use a plastic welding adhesive to bond the edges of the crack together. Carefully inject the adhesive into the crack, ensuring that it penetrates through the full depth of the damage. Use a plastic spreader or a gloved finger to smooth the adhesive and remove any excess.

c. Fill the damaged area with repair material: Once the reinforcement mesh or welding adhesive has been applied, fill the damaged area with a compatible plastic repair material, such as a two-part epoxy filler or thermoplastic repair compound. Slightly overfill the area to allow for sanding and shaping.

d. Allow the repair material to cure: Let the repair material cure fully according to the manufacturer's instructions before proceeding with shaping and sanding.

Shaping, Sanding, and Finishing the Repaired Area

After the reinforced and filled area has cured, follow the same shaping, sanding, and finishing steps outlined in the previous section on repairing plastic bumpers.

a. Shape and sand the repaired area: Use progressively finer grits of sandpaper (80-grit to 400-grit or finer) to shape and smooth the repaired area, matching the original contours of the bumper and feathering the edges to create a seamless transition.

b. Prime and paint the repaired area: Apply a compatible primer and paint to the repaired area, extending slightly beyond the repair to ensure a smooth blend with the surrounding bumper. Use a spray gun or aerosol can to apply the primer and paint in even, overlapping passes, allowing each coat to dry before applying the next.

c. Clear coat and buff (optional): If desired, apply a compatible clear coat over the repaired area and buff it to a smooth, glossy finish using a fine-grit polishing compound and a buffing pad.

Considerations for Severe Damage or Partial Replacement

In some cases, the extent of the cracks or breaks may be too severe for a reliable repair, or the cost of the repair may exceed the cost of partial or complete bumper replacement.

In these situations, it may be necessary to consider alternative options.

a. Partial bumper replacement: If the cracks or breaks are confined to a specific area of the bumper, such as a corner or a small section, it may be possible to replace just that portion of the bumper rather than the entire assembly. This can be a cost-effective solution for localized, severe damage.

b. Complete bumper replacement: If the cracks or breaks are extensive, span multiple areas of the bumper, or compromise its structural integrity, complete bumper replacement may be the most appropriate solution. This involves removing the entire damaged bumper and installing a new or used replacement bumper that matches the vehicle's make, model, and color.

When considering partial or complete bumper replacement, be sure to factor in the availability and cost of the replacement parts, as well as the labor required for removal and installation. In some cases, the cost of replacement may be comparable to or even less than the cost of an extensive repair, making it a more practical and economical choice.

By carefully assessing the severity of the cracks or breaks, employing appropriate reinforcement and filling techniques, and following best practices for shaping, sanding, and finishing, it is possible to successfully repair many types of cracked or broken bumpers. However, it's essential to use good judgment and consider the limitations of repair when dealing with severe damage, always prioritizing the structural integrity and safety of the repaired bumper.

Replacing Irreparable Bumpers

In some cases, a bumper may be too severely damaged to be repaired effectively, or the cost of repair may exceed the cost of replacement. In these situations, replacing the irreparable bumper with a new or used replacement is often the most practical and economical solution. In this section, we'll explore the key steps and considerations involved in replacing irreparable bumpers, from selecting the right replacement to installing and finishing the new bumper.

Choosing the Right Replacement Bumper
The first step in replacing an irreparable bumper is to select the appropriate replacement bumper for the specific make, model, and year of the vehicle.

Factors to consider when choosing a replacement bumper:
a. OEM vs. aftermarket: Decide whether to use an original equipment manufacturer (OEM) bumper or an aftermarket replacement. OEM bumpers are identical to the factory-installed bumper and ensure a perfect fit and finish, but they can be more expensive. Aftermarket bumpers are often less costly but may require additional fitting or modification.

b. Material and construction: Consider the material and construction of the replacement bumper. Most modern bumpers are made from thermoplastic materials, such as polypropylene or polyurethane, but some aftermarket bumpers may use different materials or construction methods that could affect the bumper's durability, flexibility, or ease of installation.

c. Primed vs. painted: Determine whether to purchase a primed or pre-painted replacement bumper. Primed bumpers require painting to match the vehicle's color, while pre-painted bumpers come finished in the correct color code for the vehicle. Pre-painted bumpers can save time and effort, but they may be more expensive and may not match the vehicle's paint perfectly.

d. New vs. used: Consider whether to purchase a new or used replacement bumper. New bumpers offer the best fit, finish, and durability, but they can be costly. Used bumpers, sourced from salvage yards or online marketplaces, can be a more budget-friendly option but may require additional cleaning, preparation, or repair before installation.

Removing the Damaged Bumper
Before installing the replacement bumper, the damaged bumper must be carefully removed from the vehicle.

Steps for removing the damaged bumper:
a. Disconnect any electrical components: If the bumper features any electrical components, such as fog lights, parking sensors, or wiring harnesses, disconnect these components and set them aside for later reconnection.

b. Remove any fasteners or clips: Using the appropriate tools, remove any bolts, screws, or plastic clips that secure the bumper to the vehicle's frame or fenders. These fasteners may be located along the top, sides, or bottom edges of the bumper.

c. Detach the bumper from the vehicle: Carefully pull the bumper away from the vehicle, being mindful of any remaining fasteners or clips that may need to be released. If the bumper is difficult to remove, consult the vehicle's service manual for specific instructions or guidance.

d. Remove any attached components: If the damaged bumper has any components that need to be transferred to the replacement bumper, such as mounting brackets, fog light housings, or license plate frames, remove these components and set them aside.

Preparing the Replacement Bumper
Before installing the replacement bumper, it may need to be prepared or modified to ensure a proper fit and finish.

Steps for preparing the replacement bumper:
a. Clean and inspect the bumper: If using a used or aftermarket replacement bumper, thoroughly clean the bumper and inspect it for any damage, defects, or missing components. Repair or replace any issues before proceeding with installation.

b. Paint the bumper (if necessary): If the replacement bumper is primed or unpainted, paint it to match the vehicle's color code using the same techniques outlined in the previous sections on bumper repair and painting. Allow the paint to fully cure before handling or installing the bumper.

c. Transfer any components: If any components need to be transferred from the damaged bumper to the replacement bumper, such as mounting brackets or fog light housings, install these components using the appropriate fasteners and techniques.

d. Modify the bumper (if necessary): In some cases, particularly with aftermarket bumpers, minor modifications may be necessary to ensure a proper fit. This could include trimming excess material, enlarging mounting holes, or adjusting the shape of the bumper to match the vehicle's contours. Consult the bumper's installation instructions or a professional for guidance on any necessary modifications.

Installing the Replacement Bumper
With the damaged bumper removed and the replacement bumper prepared, the new bumper can be installed on the vehicle.

Steps for installing the replacement bumper:
a. Position the bumper on the vehicle: Carefully align the replacement bumper with the vehicle's frame and fenders, ensuring that it is centered and level. If necessary, enlist the help of an assistant to hold the bumper in place while you secure it.

b. Secure the bumper with fasteners: Using the original or supplied fasteners, attach the bumper to the vehicle's frame or fenders, starting with the central mounting points and working outward. Ensure that each fastener is tightened to the appropriate torque specification to prevent loosening or damage.

c. Reconnect any electrical components: If the bumper features any electrical components, such as fog lights or parking sensors, reconnect the wiring harnesses and test the components to ensure proper functionality.

d. Adjust the bumper alignment: If necessary, make any final adjustments to the bumper's alignment, ensuring that it is flush with the surrounding body panels and that any gaps or seams are even and consistent. Some bumpers may have slotted mounting holes or adjustable brackets that allow for fine-tuning of the alignment.

Finishing and Quality Control
After installing the replacement bumper, it's essential to perform a final inspection and quality control check to ensure that the bumper is properly installed and finished.

Steps for finishing and quality control:
a. Check the fit and alignment: Carefully examine the installed bumper from all angles, checking for any gaps, misalignments, or inconsistencies in the fit. Make any necessary adjustments to ensure a seamless, factory-like appearance.

b. Inspect the paint and finish: If the replacement bumper was painted, closely inspect the paint for any defects, such as runs, sags, or orange peel. Touch up any imperfections using the appropriate techniques and materials.

c. Test any electrical components: If the bumper features any electrical components, thoroughly test their functionality to ensure that they operate correctly and consistently.

c. Reconnect any electrical components: If the bumper features any electrical components, such as fog lights or parking sensors, reconnect the wiring harnesses and test the components to ensure proper functionality.

d. Adjust the bumper alignment: If necessary, make any final adjustments to the bumper's alignment, ensuring that it is flush with the surrounding body panels and that any gaps or seams are even and consistent. Some bumpers may have slotted mounting holes or adjustable brackets that allow for fine-tuning of the alignment.

Finishing and Quality Control
After installing the replacement bumper, it's essential to perform a final inspection and quality control check to ensure that the bumper is properly installed and finished.

Steps for finishing and quality control:
a. Check the fit and alignment: Carefully examine the installed bumper from all angles, checking for any gaps, misalignments, or inconsistencies in the fit. Make any necessary adjustments to ensure a seamless, factory-like appearance.

b. Inspect the paint and finish: If the replacement bumper was painted, closely inspect the paint for any defects, such as runs, sags, or orange peel. Touch up any imperfections using the appropriate techniques and materials.

c. Test any electrical components: If the bumper features any electrical components, thoroughly test their functionality to ensure that they operate correctly and consistently.

d. Clean and detail the bumper: Finally, clean and detail the newly installed bumper to remove any dirt, fingerprints, or residue from the installation process. Apply any necessary protective coatings, such as wax or ceramic coating, to help preserve the finish and protect against environmental damage.

By carefully selecting the right replacement bumper, following proper removal and installation techniques, and conducting thorough quality control, you can successfully replace irreparable bumpers and restore a vehicle's appearance and functionality. While bumper replacement can be a more involved and costly process than repair, it is often the most effective solution for severely damaged or degraded bumpers, ensuring the best possible outcome for the vehicle and the customer.

Painting and Finishing Bumpers

Whether you're repairing a damaged bumper or installing a new, primed replacement, painting and finishing the bumper is a critical step in restoring its appearance and protecting it from the elements. A well-executed paint job can make the difference between a seamless, factory-like finish and an obvious, unsightly repair. In this section, we'll explore the key steps and techniques involved in painting and finishing bumpers, from surface preparation to final detailing.

Preparing the Bumper for Paint
Before painting a bumper, it's essential to properly prepare the surface to ensure optimal adhesion, durability, and finish quality.

Steps for preparing a bumper for paint:
a. Clean the bumper: Thoroughly wash the bumper using a mild detergent and water to remove any dirt, grime, or contaminants. If the bumper has been previously repaired or sanded, use a plastic cleaner or degreaser to remove any residue or oils that may interfere with paint adhesion.

b. Sand the bumper: Using progressively finer grits of sandpaper (320-grit to 600-grit), lightly sand the entire bumper to create a smooth, even surface for painting. If the bumper has been repaired or filled, focus on blending the repaired areas with the surrounding surface, ensuring a seamless transition.

c. Apply a plastic adhesion promoter: If the bumper is made of a non-porous plastic material, apply a plastic adhesion promoter to help the primer and paint adhere properly. Follow the manufacturer's instructions for application and drying times.

d. Mask off surrounding areas: Using high-quality masking tape and paper or plastic sheeting, carefully mask off any areas of the vehicle that you don't want to be painted, such as the headlights, grille, or fenders. Make sure the masking is secure and free of gaps or wrinkles that could allow overspray to penetrate.

Priming the Bumper
Applying a high-quality primer to the bumper is crucial for creating a stable, uniform base for the topcoat and helping to prevent chips, peeling, or other paint failures.

Steps for priming a bumper:
a. Select the appropriate primer: Choose a primer that is compatible with the bumper material and the type of paint you will be using. For most plastic bumpers, a flexible, high-build primer is recommended to help fill minor imperfections and promote adhesion.

b. Apply the primer: Using a spray gun or aerosol can, apply the primer to the bumper in thin, even coats, overlapping each pass by about 50% to ensure consistent coverage. Follow the manufacturer's instructions for spraying distance, pressure, and flash times between coats.

c. Allow the primer to dry: Let the primer dry and cure according to the manufacturer's guidelines, which may range from 30 minutes to several hours depending on the product and ambient conditions.

d. Sand the primed surface: Once the primer has fully dried, lightly sand the surface with 600-grit or finer sandpaper to remove any imperfections and create a smooth base for the topcoat. Use a tack cloth to remove any sanding dust before proceeding.

Applying the Base Coat
The base coat is the actual color coat of the paint job, providing the bumper with its final hue and opacity.

Steps for applying the base coat:
a. Match the paint color: If the bumper is being painted to match the rest of the vehicle, use the vehicle's color code (usually found on a sticker or placard in the door jamb or under the hood) to obtain the correct paint formula from a professional automotive paint supplier.

b. Mix and strain the paint: Following the paint manufacturer's instructions, mix the paint components (base color, reducer, and hardener) in the proper ratios and strain the mixture through a fine-mesh paint strainer to remove any lumps or debris.

c. Apply the base coat: Using a spray gun, apply the base coat to the bumper in thin, even layers, allowing each coat to flash off (become matte and dry to the touch) before applying the next.

Typically, 2-4 coats are necessary to achieve full coverage and color depth, depending on the paint formula and application technique.

d. Allow the base coat to dry: Once the final coat of base color has been applied, allow it to dry and cure according to the paint manufacturer's guidelines. This may take anywhere from a few minutes to several hours, depending on the product and ambient conditions.

Applying the Clear Coat
The clear coat is a transparent layer of paint that is applied over the base coat to provide gloss, depth, and protection from UV rays, chemicals, and abrasion.

Steps for applying the clear coat:
a. Mix and strain the clear coat: Following the paint manufacturer's instructions, mix the clear coat components (clear, reducer, and hardener) in the proper ratios and strain the mixture through a fine-mesh paint strainer.

b. Apply the clear coat: Using a spray gun, apply the clear coat to the bumper in thin, even layers, allowing each coat to flash off before applying the next. Typically, 2-3 coats of clear are necessary to achieve optimal gloss, depth, and protection.

c. Allow the clear coat to cure: After applying the final coat of clear, allow it to cure according to the paint manufacturer's guidelines. Depending on the product and ambient conditions, this may take anywhere from several hours to several days.

Color Sanding and Polishing (Optional)

If the painted bumper has minor imperfections, such as dust nibs, orange peel, or runs, color sanding and polishing can help to refine the finish and achieve a smoother, glossier result.

Steps for color sanding and polishing:

a. Allow the clear coat to fully cure: Before color sanding and polishing, ensure that the clear coat has fully cured according to the paint manufacturer's instructions. Attempting to sand or polish too soon can cause irreversible damage to the paint.

b. Wet sand the surface: Using a sanding block and progressively finer grits of wet sandpaper (1000-grit to 3000-grit), gently sand the surface of the clear coat to remove any imperfections and level out the finish. Keep the surface wet with a spray bottle or sponge to lubricate the sandpaper and prevent deep scratches.

c. Polish the surface: Using a high-quality polishing compound and a foam pad or microfiber towel, polish the sanded surface to restore gloss and remove any sanding marks. Work in small sections, using a circular motion and light to moderate pressure.

d. Refine the polish: If necessary, use a finer polishing compound or a finishing wax to further refine the gloss and remove any residual swirl marks or haze.

Reassembly and Quality Control
Once the painted bumper has fully cured and been polished (if desired), it can be reinstalled on the vehicle and given a final quality control inspection.

Steps for reassembly and quality control:
a. Reinstall the bumper: Carefully align the painted bumper with the vehicle's mounting points and secure it using the appropriate fasteners and techniques, as described in the previous section on bumper replacement.

b. Reconnect any electrical components: If the bumper features any electrical components, such as fog lights or parking sensors, reconnect the wiring harnesses and test the components to ensure proper functionality.

c. Inspect the paint and finish: Under good lighting conditions, closely examine the painted bumper for any defects, such as runs, sags, orange peel, or color mismatches. If necessary, make any touch-ups or corrections using the appropriate techniques and materials.

d. Clean and detail the bumper: Finally, thoroughly clean and detail the freshly painted bumper to remove any dust, residue, or fingerprints from the installation process. Apply a high-quality wax or sealant to help protect the finish and enhance the gloss.

By following these painting and finishing techniques, and paying close attention to surface preparation, product selection, and application methods, you can achieve a professional-quality paint job that seamlessly blends the repaired or replaced bumper with the rest of the vehicle. With practice and patience, your bumper paint work will rival that of a factory finish, ensuring customer satisfaction and a long-lasting, durable result.

Chapter 5
Hail Damage Repair
Understanding Hail Damage

Hail damage is a common and often severe type of vehicle damage caused by the impact of hailstones during a hailstorm. Hailstones can vary in size from a small pea to a large grapefruit, and the larger the hailstone, the more extensive the damage it can cause to a vehicle's exterior. In this section, we'll explore the characteristics of hail damage, the factors that influence its severity, and the common areas of a vehicle most susceptible to hail damage.

Characteristics of Hail Damage
Hail damage on vehicles typically manifests as a series of small, round dents or dimples on the exterior surfaces, such as the hood, roof, trunk, and body panels.

Common characteristics of hail damage include:
a. Dent size: Hail dents can range in size from a few millimeters to several inches in diameter, depending on the size of the hailstones and the force of their impact.

b. Dent depth: The depth of hail dents can vary from shallow to deep, with more severe damage often resulting in creases or folds in the metal.

c. Dent shape: Hail dents are typically round or oval in shape, reflecting the shape of the hailstones that caused them. However, irregularly shaped dents can occur if the hailstone strikes the surface at an angle or if multiple hailstones strike the same area.

d. Paint condition: In most cases, hail damage does not directly affect the vehicle's paint, as the dents are caused by blunt force trauma rather than sharp impacts. However, severe hail damage can cause the paint to crack or chip, especially on areas with sharp creases or edges.

Factors Influencing Hail Damage Severity
The severity of hail damage on a vehicle can vary depending on several factors, including:

a. Hailstone size: Larger hailstones generally cause more severe damage than smaller ones, as they have greater mass and kinetic energy upon impact.

b. Wind speed and direction: Strong winds can accelerate hailstones, increasing their impact force and resulting in more severe damage. The direction of the wind can also influence which areas of the vehicle are most affected.

c. Vehicle location and orientation: Vehicles that are parked in the open or are directly exposed to falling hailstones are more likely to sustain severe damage than those that are parked under cover or are partially shielded by buildings or trees. The orientation of the vehicle relative to the storm can also affect the distribution and severity of the damage.

d. Material and thickness of the vehicle's panels: The type of material and the thickness of the vehicle's exterior panels can influence the extent of hail damage. Thicker, more rigid materials, such as steel, are generally more resistant to denting than thinner, more flexible materials, such as aluminum or plastic.

Common Areas Affected by Hail Damage
While hail damage can occur on any exposed surface of a vehicle, some areas are more susceptible to damage than others.

Common areas affected by hail damage include:
a. Hood: The hood is often one of the most severely damaged areas during a hailstorm, as it is large, flat, and directly exposed to falling hailstones.

b. Roof: Like the hood, the roof is a large, flat surface that is prone to extensive hail damage, especially on vehicles with thinner or less rigid roof panels.

c. Trunk: The trunk lid, being a horizontal surface, is also highly susceptible to hail damage, particularly on sedans and coupes.

d. Body panels: The vehicle's fenders, doors, and quarter panels can also sustain hail damage, although the severity may vary depending on the angle and force of the hailstone impacts.

e. Glass and mirrors: In extreme cases, large hailstones can crack or shatter windshields, rear windows, and side mirrors, posing a safety risk and requiring immediate replacement.

Assessing Hail Damage
When evaluating a vehicle for hail damage, it's essential to perform a thorough and systematic inspection to determine the extent and severity of the damage.

Steps for assessing hail damage:
a. Inspect the vehicle in good lighting conditions: Park the vehicle in a well-lit area, preferably outdoors in natural light, to ensure that all dents and imperfections are clearly visible.

b. Check all exposed surfaces: Systematically examine the hood, roof, trunk, body panels, and glass for signs of hail damage, using a flashlight or a dent light to highlight any dents or irregularities.

c. Estimate the number and size of dents: Count the number of dents on each panel and note their approximate sizes, as this information will be essential for determining the best repair approach and estimating costs.

d. Look for paint damage: Carefully inspect the dented areas for any signs of cracked, chipped, or flaking paint, as this may require additional repair steps beyond dent removal.

e. Document the damage: Take clear, well-lit photographs of the hail-damaged areas from multiple angles to document the extent and severity of the damage. These photos can be used for insurance claims, repair estimates, and before-and-after comparisons.

By understanding the characteristics, factors, and common areas affected by hail damage, as well as the steps for properly assessing the damage, you'll be better equipped to make informed decisions about the most appropriate repair methods and to provide accurate estimates to customers. In the following sections, we'll explore various techniques for repairing hail damage, from paintless dent repair to conventional body repair methods.

Evaluating the Extent of Hail Damage

Once you have a general understanding of the characteristics and common areas affected by hail damage, the next step is to thoroughly evaluate the extent of the damage on a specific vehicle. Assessing the severity, distribution, and complexity of the hail dents will help you determine the most appropriate repair approach, estimate the time and costs involved, and communicate effectively with the customer. In this section, we'll delve into the key factors to consider when evaluating the extent of hail damage and provide a step-by-step guide for conducting a comprehensive assessment.

Factors to Consider in Hail Damage Evaluation

When evaluating the extent of hail damage on a vehicle, several key factors should be taken into account:

a. Number of dents: The total number of hail dents on the vehicle is a primary indicator of the extent of the damage. A higher number of dents generally corresponds to a more severe and time-consuming repair process.

b. Size and depth of dents: The size and depth of the individual hail dents can vary significantly, from small, shallow dings to large, deep craters. Larger and deeper dents often require more complex repair techniques and may be more challenging to restore completely.

c. Location and distribution of dents: The location and distribution of the hail dents on the vehicle can impact the repair process. Dents on flat, easily accessible panels, such as the hood or roof, are generally easier to repair than those on curved, contoured, or hard-to-reach areas, such as the fenders or quarter panels.

d. Presence of creases or sharp edges: Hail dents with creases or sharp edges can be more challenging to repair than smooth, rounded dents. These types of dents may require additional repair steps or even panel replacement in severe cases.

e. Paint condition: The condition of the paint on the hail-damaged areas is another crucial factor to consider. If the paint has cracked, chipped, or flaked due to the hail impact, it may require touch-up or repainting in addition to dent repair.

Step-by-Step Guide for Evaluating Hail Damage
To conduct a thorough and accurate evaluation of hail damage, follow these steps:

a. Perform a visual inspection:
- Park the vehicle in a well-lit area, preferably outdoors in natural light.
- Systematically examine all exposed surfaces of the vehicle, including the hood, roof, trunk, fenders, doors, and quarter panels.
- Use a flashlight or dent light to highlight any dents, dings, or irregularities in the surface.
- Pay close attention to areas that are prone to severe hail damage, such as the hood, roof, and trunk.

b. Count and document the dents:
- Count the total number of hail dents on each panel and the vehicle as a whole.
- Make note of the size and depth of the dents, categorizing them as small, medium, or large.

- Record the location and distribution of the dents on each panel, noting any clusters or patterns.
- Use a hail damage diagram or a vehicle outline to visually map the location and severity of the dents.

c. Assess paint condition:
- Carefully inspect the hail-damaged areas for any signs of paint damage, such as cracks, chips, or flaking.
- Determine the extent of the paint damage and whether it will require touch-up or repainting.
- Make note of any pre-existing paint damage or imperfections that may affect the repair process or outcome.

d. Check for additional damage:
- Examine the vehicle for any additional damage that may have occurred during the hailstorm, such as broken glass, damaged mirrors, or bent antenna.
- Determine whether any of these components will need to be repaired or replaced in addition to the hail dent repair.

e. Document the evaluation:
- Take clear, well-lit photographs of the hail damage from multiple angles, ensuring that all affected panels are captured.
- Record your findings, including the total number of dents, their sizes, locations, and any paint or additional damage, in a detailed damage report.
- Use this documentation for insurance claims, repair estimates, and communication with the customer.

Determining the Repair Approach
Based on your evaluation of the extent of the hail damage, you can determine the most appropriate repair approach for the vehicle.

Factors to consider when selecting a repair approach:
a. Severity of the damage: If the hail damage is relatively minor, with a low number of small, shallow dents, paintless dent repair (PDR) may be the most efficient and cost-effective approach. However, if the damage is severe, with numerous large, deep dents or creases, conventional body repair techniques may be necessary.

b. Location and accessibility of the dents: PDR is generally most effective on dents located on flat, easily accessible panels, such as the hood, roof, or trunk. Dents on curved or contoured areas, or those in hard-to-reach locations, may require conventional repair methods or even panel replacement.

c. Paint condition: If the paint on the hail-damaged areas is intact and in good condition, PDR can often be performed without the need for touch-up or repainting. However, if the paint is cracked, chipped, or flaking, conventional repair techniques, including filling, sanding, and repainting, may be required.

d. Customer preferences and budget: The customer's preferences and budget should also be taken into account when determining the repair approach. Some customers may prioritize a quick and affordable repair, even if it means some minor imperfections remain, while others may insist on a flawless, showroom-quality finish, regardless of the cost or time involved.

By carefully evaluating the extent of the hail damage and considering these factors, you can select the most appropriate repair approach for each individual case, balancing the needs of the vehicle, the customer, and your shop's capabilities. In the following sections, we'll explore specific PDR techniques for hail damage repair and discuss scenarios where conventional repair methods may be necessary.

PDR Techniques for Hail Damage Repair

Paintless Dent Repair (PDR) is often the preferred method for repairing hail damage on vehicles, as it is less invasive, more cost-effective, and faster than conventional body repair techniques. PDR involves using specialized tools to manipulate and massage the metal from the backside of the panel, gently pushing the dents out without the need for fillers, sanding, or repainting. In this section, we'll explore the specific PDR techniques used for hail damage repair, the advantages and limitations of PDR in this context, and the step-by-step process for performing PDR on hail-damaged panels.

Advantages of PDR for Hail Damage Repair

PDR offers several key advantages over conventional body repair methods when it comes to repairing hail damage:

a. Preserves the vehicle's original paint: Since PDR does not require sanding, filling, or repainting, it maintains the integrity of the vehicle's factory paint job, helping to retain its value and appearance.

b. Less expensive than conventional repairs: PDR is typically more affordable than conventional body repair, as it requires fewer materials, less labor, and no paint-related costs.

c. Faster turnaround time: PDR can often be completed in a matter of days, depending on the extent of the damage, whereas conventional repairs may take several weeks due to the multiple stages involved, such as filling, sanding, priming, painting, and curing.

d. Environmentally friendly: PDR minimizes the use of chemicals, solvents, and fillers, making it a more eco-friendly option compared to conventional repair methods.

Limitations of PDR for Hail Damage Repair

While PDR is highly effective for repairing many types of hail damage, it does have some limitations:

a. Severe dents or creases: PDR may not be suitable for repairing very large, deep, or sharp-edged dents, as the metal may have stretched or become too distorted to restore fully.

b. Paint damage: If the hail impact has caused the paint to crack, chip, or flake, PDR alone will not address these issues, and conventional paint repair techniques may be necessary.

c. Hard-to-reach areas: Some hail dents may be located in areas that are difficult to access from the backside of the panel, such as near the edges or on heavily contoured surfaces, limiting the effectiveness of PDR.

PDR Tools and Equipment for Hail Damage Repair

To perform PDR on hail-damaged panels, technicians use a variety of specialized tools and equipment, including:

a. Dent lights: High-intensity LED lights that highlight the contours and shadows of the dents, helping technicians to visualize the damage and monitor progress during the repair process.

b. Dent pushing tools: Long, slender metal rods with various tip shapes and sizes that are used to apply pressure to the backside of the dent, gradually pushing it out.

c. Glue pulling systems: Adhesive-based tools that involve attaching small tabs or "buttons" to the front side of the dent, then using a bridge or slide hammer to gently pull the dent out.

d. Plastic and nylon tools: Soft, non-marring tools used for applying pressure to the dent without the risk of damaging the paint or leaving tool marks.

e. Magnets and suction cups: Devices used to secure dent lights or other tools to the panel, freeing up the technician's hands for more precise dent manipulation.

Step-by-Step PDR Process for Hail Damage Repair
The PDR process for repairing hail damage typically involves the following steps:
a. Assess the damage:
- Thoroughly inspect the hail-damaged panels, noting the number, size, depth, and location of the dents.
- Determine whether PDR is a suitable repair method for the specific damage and the panel's accessibility.

b. Gain access to the backside of the panel:
- Carefully remove any interior trim, insulation, or other components that obstruct access to the backside of the dent.
- Use protective cloths or pads to prevent scratches or damage to the vehicle's interior.

c. Apply pressure to the dent:
- Using a dent light to illuminate the damage, select the appropriate dent pushing tool for the size and shape of the dent.
- Apply gentle, consistent pressure to the backside of the dent, working from the outer edges towards the center.
- Continuously monitor progress using the dent light, adjusting the angle and pressure of the tool as needed.

d. Refine the repair:
- As the dent becomes shallower, switch to finer, more precise dent pushing tools to avoid over-correction or creating high spots.
- If necessary, use glue pulling or tapping techniques to address any remaining imperfections or shallow dents.
- Ensure that the repaired area is level with the surrounding panel and free of any visible distortions.

e. Clean up and reassemble:
- Carefully remove any interior components, protective cloths, or tools from the work area.
- Reinstall any trim pieces or insulation that were removed during the repair process.
- Thoroughly clean and inspect the repaired panel to ensure a seamless, high-quality finish.

Best Practices for PDR on Hail-Damaged Panels
To achieve the best possible results when performing PDR on hail-damaged panels, keep these best practices in mind:

a. Work in a well-lit environment: Ensure that your workspace has ample lighting, including high-quality dent lights, to help you accurately assess and repair the damage.

b. Use the right tools for the job: Select dent pushing tools that are appropriate for the size, shape, and location of each dent, and use non-marring tools or protective covers to avoid damaging the paint.

c. Take your time: Work slowly and methodically, applying gentle, controlled pressure to the dent to avoid over-correction or creating new distortions.

d. Keep the panel clean: Regularly wipe away any dirt, dust, or debris that may accumulate on the panel during the repair process to maintain a clear view of the damage and avoid contaminating the tools.

e. Know when to stop: If a dent proves too challenging to repair fully using PDR, know when to stop and consider alternative repair methods, such as conventional body filler or panel replacement, to avoid causing further damage or wasting time.

By mastering these PDR techniques and best practices for hail damage repair, you can provide your customers with a cost-effective, efficient, and high-quality solution for restoring their hail-damaged vehicles. While PDR may not be suitable for every hail damage scenario, it remains the preferred method for the majority of cases, offering significant advantages over conventional repair techniques.

Conventional Repair Methods for Severe Hail Damage

While Paintless Dent Repair (PDR) is the preferred method for addressing most hail damage, there are instances where the damage is too severe or complex for PDR alone. In these cases, conventional repair methods, such as body filler application, sanding, and repainting, may be necessary to restore the vehicle's exterior to its pre-damage condition. In this section, we'll explore the scenarios where conventional repair methods are typically required, the tools and materials used, and the step-by-step process for performing these repairs on severely hail-damaged panels.

Scenarios Requiring Conventional Repair Methods
Conventional repair methods are typically necessary for hail damage in the following situations:

a. Deep, sharp-edged dents: If the hail damage includes very deep dents with sharp edges or creases, the metal may be stretched or distorted beyond the point where PDR can fully restore it. In these cases, body filler may be needed to fill in the damaged area and create a smooth, even surface.

b. Extensive paint damage: If the hail impact has caused significant paint damage, such as large chips, cracks, or flaking, PDR alone will not address these issues. Conventional repair methods, including sanding, filling, priming, and repainting, will be necessary to restore the paint's integrity and appearance.

c. Inaccessible dents: Some hail dents may be located in areas that are difficult or impossible to access from the backside of the panel, such as near the edges, on heavily contoured surfaces, or in tight corners. In these cases, conventional repair methods may be the only option for restoring the damaged area.

d. Combination of PDR and conventional repairs: In some instances, a combination of PDR and conventional repair techniques may be used to address different aspects of the hail damage. For example, shallow dents may be repaired using PDR, while deeper dents or areas with paint damage may require conventional methods.

Tools and Materials for Conventional Hail Damage Repair
To perform conventional repairs on severely hail-damaged panels, technicians use a variety of tools and materials, including:

a. Body filler: A two-part polyester resin used to fill in deep dents, creases, and other surface imperfections. Body filler is mixed with a hardener and applied to the damaged area, then sanded smooth once cured.

b. Sandpaper: Various grits of sandpaper, ranging from coarse (80-grit) to fine (400-grit or higher), are used to shape and smooth the body filler, as well as to prepare the surrounding paint for refinishing.

c. Primer: A specialized paint product used to create a uniform surface for the topcoat and promote proper adhesion.

Primer is available in various formulations, such as epoxy, urethane, or self-etching, depending on the substrate and desired properties.

d. Paint: Automotive paint, typically a basecoat/clearcoat system, is used to refinish the repaired area and blend it with the surrounding paint. The basecoat provides the color, while the clearcoat adds depth, gloss, and protection.

e. HVLP spray gun: A high-volume, low-pressure spray gun is used to apply the primer, basecoat, and clearcoat in thin, even layers, minimizing overspray and ensuring a smooth, professional finish.

f. Putty knives and spreaders: Flexible metal or plastic tools used to mix and apply the body filler, as well as to shape and smooth it before sanding.

g. Sanding blocks and boards: Rigid backing materials used to support the sandpaper and ensure an even, consistent sanding surface, preventing the creation of low spots or uneven areas.

Step-by-Step Conventional Repair Process for Severe Hail Damage

The conventional repair process for severe hail damage typically involves the following steps:

a. **Assess and document the damage:**
- Thoroughly inspect the hail-damaged panels, noting the location, size, depth, and severity of each dent, as well as any associated paint damage.
- Take detailed photos and measurements of the damage for documentation and insurance purposes.

b. **Prepare the damaged area:**
- Clean the panel thoroughly to remove any dirt, debris, or contaminants that could interfere with the repair process.
- Sand the damaged area and surrounding paint to create a rough, feathered edge that will promote proper adhesion of the body filler and new paint.

c. **Apply and shape the body filler:**
- Mix the body filler and hardener according to the manufacturer's instructions, then apply it to the damaged area using a putty knife or spreader.
- Overfill the dent slightly to allow for sanding and shaping, then let the filler cure fully (typically 15-30 minutes, depending on the product).
- Once cured, sand the body filler using progressively finer grits of sandpaper (80-grit to 150-grit) to shape and smooth it, creating a seamless transition with the surrounding panel.

d. **Prime and block sand:**
- Apply a high-quality primer to the repaired area, extending slightly beyond the body filler to ensure proper adhesion and blending.

- Once the primer has dried, block sand it using fine-grit sandpaper (220-grit to 400-grit) to create an ultra-smooth surface for the topcoat.

e. Refinish and blend the paint:
- Using an HVLP spray gun, apply the basecoat to the primed area, matching the original paint color as closely as possible.
- Apply the clearcoat over the basecoat, extending slightly beyond the repair area to ensure a smooth, uniform finish.
- Blend the new paint into the surrounding area using various techniques, such as a soft-edge blending, to create a seamless, invisible repair.

f. Polish and detail:
- Once the paint has cured fully (typically 24-48 hours), polish the repaired area using progressively finer polishing compounds to remove any minor imperfections and restore gloss.
- Clean and detail the entire vehicle to ensure a consistent, high-quality appearance.

Best Practices for Conventional Hail Damage Repair
To achieve the best possible results when performing conventional repairs on severely hail-damaged panels, keep these best practices in mind:

a. Use high-quality materials: Invest in premium body filler, primer, paint, and sandpaper to ensure optimal durability, adhesion, and finish quality.

b. Work in a clean, controlled environment: Perform repairs in a well-ventilated, dust-free area to minimize contaminants and ensure a smooth, even finish.

c. Allow adequate curing time: Follow the manufacturer's guidelines for curing times on body filler, primer, and paint to ensure proper hardness and adhesion.

d. Use proper technique: Apply body filler, primer, and paint using smooth, even strokes, and sand in a consistent, methodical manner to avoid creating uneven areas or visible repair marks.

e. Blend the repair seamlessly: Use advanced blending techniques and high-quality paint products to ensure that the repaired area is virtually indistinguishable from the surrounding paint.

While conventional repair methods for severe hail damage are more time-consuming and labor-intensive than PDR, they remain an essential aspect of comprehensive hail damage repair. By mastering these techniques and best practices, you can provide your customers with a full range of repair options and ensure that even the most challenging hail damage can be addressed effectively.

Insurance Claims and Hail Damage Repair

Hail damage is one of the most common types of automotive insurance claims, and navigating the claims process can be complex and time-consuming for both vehicle owners and repair professionals. Understanding the ins and outs of insurance claims related to hail damage repair is crucial for ensuring that the repair process goes smoothly, customers receive fair compensation, and your shop is paid appropriately for its work. In this section, we'll explore the key aspects of insurance claims for hail damage repair, including the claims process, working with insurance adjusters, and best practices for documentation and communication.

The Hail Damage Insurance Claim Process
The typical insurance claim process for hail damage involves several key steps:

a. Policyholder initiates the claim:
- The vehicle owner contacts their insurance company to report the hail damage and initiate the claim process.
- The policyholder provides basic information about the damage, including when and where it occurred, and the extent of the damage.

b. Insurance company assigns an adjuster:
- The insurance company assigns a claims adjuster to assess the damage and determine the cost of repair.
- The adjuster may be an employee of the insurance company or an independent contractor.

c. Adjuster inspects the damage:
- The adjuster schedules an appointment to inspect the hail-damaged vehicle in person.
- During the inspection, the adjuster documents the extent of the damage, takes photographs, and estimates the cost of repair based on their assessment.

d. Repair shop provides an estimate:
- The policyholder may choose to have their vehicle repaired at a preferred shop or one recommended by the insurance company.
- The repair shop performs a thorough inspection of the hail damage and provides a detailed repair estimate to the policyholder and the insurance company.

e. Negotiations and approval:
- The insurance adjuster reviews the repair shop's estimate and may negotiate with the shop to reach an agreed-upon price for the repairs.
- Once the estimate is approved, the insurance company authorizes the repair shop to proceed with the work.

f. Repair work is completed:
- The repair shop performs the necessary hail damage repairs, keeping the policyholder and insurance company informed of any changes or additional work required.
- Upon completion, the repair shop invoices the insurance company for the approved amount, and the policyholder pays any deductible or out-of-pocket expenses.

Working with Insurance Adjusters

Developing a positive working relationship with insurance adjusters is essential for streamlining the claims process and ensuring fair compensation for your hail damage repair work. Consider the following tips:

a. Be professional and courteous: Treat insurance adjusters with respect and maintain a professional demeanor in all interactions, whether in person, over the phone, or via email.

b. Provide thorough documentation: Supply adjusters with detailed, itemized estimates that clearly outline the scope of work, labor hours, and materials required for the repair. Include photographs and measurements to support your assessment.

c. Justify your estimate: Be prepared to explain and defend your repair estimate, providing evidence and expertise to support your recommendations.

d. Be flexible and open to negotiation: Understand that adjusters are working to balance the interests of the policyholder and the insurance company. Be willing to engage in constructive dialogue and find mutually agreeable solutions when disagreements arise.

e. Maintain open lines of communication: Keep adjusters informed of any changes, delays, or additional work required during the repair process. Prompt, transparent communication can help build trust and facilitate smoother claims resolution.

Documentation and Record-Keeping

Accurate, detailed documentation is critical for successfully navigating the insurance claims process and protecting your shop's interests. Key documentation and record-keeping practices include:

a. Detailed repair estimates: Create comprehensive repair estimates that include a clear breakdown of labor hours, materials, and associated costs. Use standardized formatting and terminology to ensure clarity and consistency.

b. Photographs and videos: Capture high-quality images and videos of the hail damage before, during, and after the repair process. This visual documentation can provide valuable evidence to support your repair recommendations and justify your work.

c. Customer authorization and agreements: Obtain signed agreements from customers authorizing the repair work and acknowledging any deductibles or out-of-pocket expenses. Keep copies of these agreements on file.

d. Insurance correspondence: Maintain records of all correspondence with insurance companies, including emails, letters, and notes from phone conversations. This documentation can be invaluable in the event of disputes or misunderstandings.

e. Invoices and payment records: Keep detailed invoices and payment records for each hail damage repair job, including any adjustments or supplemental payments from the insurance company.

Best Practices for Insurance Claims and Hail Damage Repair

To optimize the insurance claims process and ensure the best possible outcomes for your shop and your customers, consider the following best practices:

a. Educate your customers: Help policyholders understand their coverage, deductibles, and the claims process. Provide guidance and support to help them navigate the system and make informed decisions.

b. Be proactive in communication: Reach out to insurance adjusters and customers regularly to provide updates, address concerns, and maintain transparency throughout the repair process.

c. Invest in training and certifications: Pursue ongoing training and certifications in hail damage repair techniques, estimating, and insurance claims processes. This expertise can help you build credibility with insurance companies and justify your repair recommendations.

d. Develop relationships with insurance companies: Cultivate positive, professional relationships with insurance companies and adjusters in your area. Attend industry events, participate in networks, and maintain open lines of communication to foster trust and collaboration.

e. Stay current with industry trends and best practices: Keep abreast of the latest advancements in hail damage repair techniques, technology, and insurance claims processes.

Chapter 6
Rust Repair and Prevention

Identifying Rust Damage

Rust is a common and often severe problem that can affect vehicles of all ages and types, particularly those exposed to harsh weather conditions, road salt, or moisture. Identifying rust damage early and accurately is crucial for determining the most appropriate repair approach and preventing further deterioration. In this section, we'll explore the different types of rust, the common areas where rust tends to develop, and the steps for thoroughly assessing the extent and severity of rust damage on a vehicle.

Types of Rust
Rust can manifest in several different forms, each with its own characteristics and implications for repair:

a. Surface rust:
- Also known as oxidation, surface rust is the earliest stage of rust development.
- It appears as a light, powdery coating on the surface of the metal, often reddish-brown in color.
- Surface rust typically does not penetrate deeply into the metal and can often be removed with sanding or light abrasion.

b. Scale rust:
- Scale rust is a more advanced stage of rust that occurs when surface rust is left untreated.

- It appears as a thicker, flakier layer of rust that begins to penetrate the surface of the metal.
- Scale rust can often be removed through sanding or grinding, but may require more extensive repair if it has caused pitting or weakening of the metal.

c. **Penetrating rust:**
- Penetrating rust is the most severe form of rust damage, occurring when scale rust is allowed to progress unchecked.
- It appears as deep, pitted areas where the rust has eaten through the metal, often creating holes or perforations.
- Penetrating rust requires extensive repair or replacement of the affected metal, as the structural integrity of the panel may be compromised.

Common Areas for Rust Development

While rust can occur on any exposed metal surface of a vehicle, some areas are more prone to rust development than others. Common areas to inspect for rust include:

a. **Door bottoms and edges:**
- The lower portions of doors, particularly along the bottom edges, are prone to rust due to their exposure to moisture and road debris.
- Look for bubbling, flaking, or perforated metal along the door bottoms and edges.

b. **Fenders and quarter panels:**
- The fenders and quarter panels, particularly near the wheel wells, are susceptible to rust from exposure to road salt, moisture, and kicked-up debris.

- Check for rust along the edges of the fenders, around the wheel well openings, and at the junction between the fender and door.

c. Hood and trunk lid:
- The hood and trunk lid, especially along the forward edge of the hood and the rearward edge of the trunk, can develop rust from exposure to moisture and debris.
- Inspect the edges and underside of the hood and trunk lid for signs of rust or paint bubbling.

d. Frame and underbody:
- The frame and underbody of a vehicle, including the floorpans, subframe, and suspension components, are constantly exposed to moisture, salt, and debris from the road.
- Check for rust on the frame rails, crossmembers, and suspension mounting points, as well as on the underside of the floorpans and trunk floor.

e. Windshield and window surrounds:
- The metal surrounds of the windshield and windows can trap moisture and debris, leading to rust development.
- Look for rust or bubbling paint around the perimeter of the windshield and window openings.

Steps for Assessing Rust Damage

To thoroughly assess the extent and severity of rust damage on a vehicle, follow these steps:

a. **Perform a visual inspection:**
   - In a well-lit area, carefully examine the vehicle's exterior, paying close attention to the common rust-prone areas listed above.
   - Look for any signs of rust, paint bubbling, or flaking, and note the location and approximate size of each affected area.

b. **Use a magnet or a screwdriver:**
   - To help identify the presence of rust beneath the paint or in hard-to-see areas, use a small magnet or the tip of a screwdriver to gently probe the surface.
   - If the magnet does not stick or the screwdriver easily penetrates the surface, this may indicate the presence of rust or weakened metal.

c. **Check the underbody:**
   - Using a lift or jack and stands, raise the vehicle to inspect the underbody and frame for rust damage.
   - Pay close attention to the frame rails, suspension mounting points, and floorpans, as these areas are particularly prone to rust.

d. **Document the damage:**
   - Take detailed photos of each rust-affected area, including close-ups and wider shots to provide context.
   - Make notes about the location, size, and severity of each rust spot, as well as any potential challenges or considerations for repair.

e. Determine the repair approach:
- Based on the extent and severity of the rust damage, determine the most appropriate repair approach, such as sanding and refinishing for surface rust, patching or replacing metal for scale rust, or more extensive fabrication for penetrating rust.
- Consider factors such as the age and value of the vehicle, the structural integrity of the affected areas, and the cost and feasibility of repair when making your determination.

By understanding the different types of rust, knowing where to look for rust damage, and following a thorough assessment process, you can accurately identify and evaluate rust damage on a vehicle. This information will be essential for developing an effective repair plan and communicating with customers about the scope and cost of the work required. In the following sections, we'll explore specific techniques for removing rust, repairing damaged metal, and preventing future rust development.

Removing Rust from Metal Surfaces

Once you have identified and assessed the rust damage on a vehicle, the next step is to remove the rust from the affected metal surfaces. Removing rust is a critical part of the repair process, as it helps to prevent further corrosion and prepares the surface for subsequent steps, such as patching, filling, or refinishing. In this section, we'll explore various methods for removing rust from metal surfaces, including manual techniques, chemical treatments, and abrasive blasting, as well as best practices for achieving a clean, rust-free surface.

Manual Rust Removal Techniques
Manual rust removal techniques involve physically abrading or scraping the rust from the metal surface using hand tools or power tools. Common manual techniques include:

a. Sanding:
- Use sandpaper or sanding discs in progressively finer grits (40-grit to 400-grit) to remove rust and smooth the metal surface.
- For small, localized rust spots, hand sanding may be sufficient, while larger areas may require the use of a power sander or grinder.
- Be cautious not to remove too much of the surrounding metal or create deep scratches that could weaken the panel.

b. Wire brushing:
- Use a wire brush, either manual or attached to a power drill, to scrub away loose rust and scale.

- Wire brushes are particularly effective for removing rust from hard-to-reach areas, such as seams, corners, and crevices.
- Be careful not to apply too much pressure, as aggressive wire brushing can scratch or damage the underlying metal.

c. Scraping:
- For heavy, flaky rust, use a scraper or putty knife to gently lift and remove the rust scales.
- Work slowly and carefully to avoid gouging or scratching the metal surface.
- Follow up with sanding or wire brushing to remove any remaining rust and smooth the surface.

Chemical Rust Removal Methods
Chemical rust removal methods involve the use of specialized products that react with the rust to dissolve or loosen it from the metal surface. Common chemical rust removal methods include:

a. Rust converters:
- Rust converters are chemical solutions that convert rust into a stable, black compound that can be painted over.
- Apply the rust converter to the rusted surface using a brush or spray bottle, following the manufacturer's instructions for application and drying time.
- Once the converter has dried, the treated surface can be sanded smooth and primed for painting.

b. Rust removers:
- Rust removers are acidic or alkaline solutions that dissolve rust without damaging the underlying metal.
- Apply the rust remover to the affected area using a brush or cloth, allowing it to penetrate and loosen the rust.
- After the recommended dwell time, scrub the surface with a wire brush or abrasive pad to remove the loosened rust, then rinse thoroughly with water.

c. Electrolysis:
- Electrolysis is a more advanced rust removal method that uses an electric current to convert rust back into solid metal.
- The rusted part is submerged in an electrolyte solution (usually water and washing soda) and connected to the negative terminal of a battery charger, while a sacrificial anode (often a piece of scrap metal) is connected to the positive terminal.
- When the current is applied, the rust is drawn away from the part and deposited on the sacrificial anode, leaving behind clean, bare metal.

Abrasive Blasting Techniques

Abrasive blasting techniques use high-pressure streams of abrasive media to remove rust, scale, and other contaminants from metal surfaces. Common abrasive blasting techniques for rust removal include:

a. **Sand blasting:**
- Sand blasting uses a high-pressure stream of sand or silica particles to remove rust and clean the metal surface.
- This method is highly effective but requires specialized equipment and a controlled environment to contain the dust and debris.
- Sand blasting can also create a rougher surface profile, which may require additional smoothing before refinishing.

b. **Soda blasting:**
- Soda blasting uses a stream of baking soda particles to remove rust and other contaminants from metal surfaces.
- This method is gentler than sand blasting and less likely to damage the underlying metal or create a rough surface profile.
- Soda blasting is also more environmentally friendly and easier to clean up than sand blasting.

c. **Media blasting:**
- Media blasting is a broad term that encompasses the use of various abrasive media, such as glass beads, plastic particles, or walnut shells, to remove rust and clean metal surfaces.
- Different media types offer different levels of aggressiveness and surface profile, allowing for customization based on the specific needs of the project.
- Media blasting generally requires specialized equipment and a controlled environment, similar to sand blasting.

Best Practices for Rust Removal

To achieve the best results when removing rust from metal surfaces, consider the following best practices:

a. Use the appropriate method for the level of rust:
- For light, surface rust, manual sanding or wire brushing may be sufficient.
- For more severe rust or larger areas, chemical treatments or abrasive blasting may be more effective and efficient.

b. Protect surrounding areas:
- When using power tools, chemicals, or abrasive blasting, be sure to protect nearby components, trim, or glass from damage or overspray.
- Use masking tape, plastic sheeting, or specialized masking products to cover and shield adjacent areas.

c. Work in a well-ventilated area:
- Many rust removal methods generate dust, fumes, or mists that can be harmful if inhaled.
- Always work in a well-ventilated area and wear appropriate personal protective equipment, such as a respirator, goggles, and gloves.

d. Clean the surface thoroughly:
- After removing the rust, clean the metal surface thoroughly with a solvent or degreaser to remove any residue, oils, or contaminants.
- A clean surface is essential for proper adhesion of subsequent coatings, such as primer, paint, or rust-inhibiting treatments.

e. Follow up with rust prevention measures:
- Once the rust has been removed and the surface has been cleaned, it's important to apply appropriate rust prevention measures, such as primer, paint, or rust-inhibiting coatings, to protect the metal from future corrosion.
- Choose rust prevention products that are compatible with the metal type and the intended use of the vehicle or component.

By selecting the appropriate rust removal method, following best practices for surface preparation and safety, and applying effective rust prevention measures, you can successfully remove rust from metal surfaces and restore the integrity and appearance of the affected areas. In the next section, we'll explore specific techniques for treating and preventing future rust development on automotive metal surfaces.

Treating and Preventing Future Rust

After removing rust from metal surfaces, it's crucial to take steps to treat the bare metal and prevent future rust development. Without proper treatment and protection, the metal will quickly begin to rust again, undermining the time and effort invested in the repair process. In this section, we'll discuss various methods for treating bare metal after rust removal, as well as strategies for preventing future rust formation, including the use of rust-inhibiting coatings, proper surface preparation, and regular maintenance practices.

Treating Bare Metal After Rust Removal
Once the rust has been removed from a metal surface, the bare metal is exposed and vulnerable to new rust formation. To prevent this, it's essential to treat the bare metal promptly and thoroughly. Common treatment methods include:

a. Phosphoric acid etching:
- Phosphoric acid is a mild acid that helps to etch and clean bare metal surfaces, promoting better adhesion of subsequent coatings.
- Apply a phosphoric acid-based metal prep solution to the bare metal surface using a brush or spray bottle, following the manufacturer's instructions for dwell time and rinsing.
- The etched surface will have a slight grayish or whitish appearance, indicating that it's ready for priming or painting.

b. Epoxy primer:
- Epoxy primers are two-part coatings that provide excellent adhesion and corrosion resistance on bare metal surfaces.
- Mix the epoxy primer components according to the manufacturer's instructions, then apply the primer to the prepared metal surface using a brush, roller, or spray gun.
- Allow the primer to cure fully before applying any additional coatings, such as filler, paint, or clear coat.

c. Self-etching primer:
- Self-etching primers are designed to chemically etch and prime bare metal surfaces in one step, simplifying the treatment process.
- Apply the self-etching primer to the prepared metal surface using a spray gun, following the manufacturer's instructions for application and flash times.
- Once the primer has dried, it can be sanded smooth and topcoated with paint or other coatings as desired.

Rust-Inhibiting Coatings and Treatments

In addition to priming and painting, various rust-inhibiting coatings and treatments can be applied to bare metal surfaces to provide long-term protection against rust formation. Some popular options include:

a. Rust-inhibiting paint:
- Rust-inhibiting paints, also known as direct-to-metal (DTM) paints, contain special additives that help to prevent rust formation on metal surfaces.

- These paints can be applied directly to properly prepared bare metal, eliminating the need for a separate primer coat.
- Choose a rust-inhibiting paint that is compatible with the metal type and the intended use of the vehicle or component, and follow the manufacturer's instructions for application and curing.

b. Rust-inhibiting wax or sealant:
- Rust-inhibiting waxes and sealants are designed to provide a protective barrier against moisture and other corrosive elements.
- These products are typically applied to bare metal or painted surfaces using a brush, sponge, or spray applicator, then allowed to dry or cure to form a water-resistant coating.
- Rust-inhibiting waxes and sealants are often used on undercarriage components, frame rails, and other areas prone to rust and corrosion.

c. Rust-inhibiting undercoating:
- Rust-inhibiting undercoatings are thick, rubberized coatings that are applied to the underside of a vehicle to protect against rust, corrosion, and road debris.
- These coatings are typically sprayed onto the prepared metal surface using specialized equipment, then allowed to dry and cure to form a durable, flexible barrier.
- Rust-inhibiting undercoatings are particularly useful for protecting frame rails, floorpans, and other undercarriage components in harsh driving conditions.

Proper Surface Preparation Techniques

Proper surface preparation is essential for ensuring the long-term effectiveness of rust-inhibiting coatings and treatments. To prepare metal surfaces for coating, follow these best practices:

a. Clean the surface thoroughly:
- Remove all dirt, grease, oil, and other contaminants from the metal surface using a degreaser, solvent, or detergent solution.
- Pay close attention to seams, crevices, and other hard-to-reach areas where dirt and debris may accumulate.

b. Roughen the surface:
- Use sandpaper, a scuff pad, or a wire brush to lightly abrade the metal surface, creating a rougher profile that will help the coating adhere better.
- Be careful not to remove too much material or create deep scratches that could compromise the strength or appearance of the metal.

c. Remove dust and debris:
- After sanding or abrading the surface, use compressed air, a tack cloth, or a vacuum to remove any dust, debris, or loose particles that could interfere with coating adhesion.
- Wipe the surface with a solvent-dampened cloth to remove any remaining residue and ensure a clean, dry surface for coating application.

Regular Maintenance and Inspection
Even with the best rust prevention measures in place, regular maintenance and inspection are critical for catching and addressing any new rust development early. To keep rust at bay, consider the following maintenance practices:

a. Wash and dry the vehicle regularly:
- Regularly washing and drying the vehicle helps to remove dirt, salt, and other contaminants that can promote rust formation.
- Pay extra attention to wheel wells, undercarriage components, and other areas where dirt and debris may collect.

b. Inspect for rust periodically:
- Perform a thorough visual inspection of the vehicle's body, frame, and undercarriage at least once or twice a year, looking for any signs of new rust development.
- Use a flashlight and mirror to check hard-to-see areas, such as the inside of frame rails and behind body panels.
- If you spot any new rust, address it promptly using the appropriate removal and treatment techniques.

c. Touch up chips and scratches:
- Any damage to the vehicle's paint or coatings can expose the bare metal beneath, creating an entry point for rust.
- Regularly inspect the vehicle's exterior for chips, scratches, or other damage, and touch up these areas with paint or rust-inhibiting coatings as needed.
- Use a fine brush or touch-up pen to apply the coating, then blend the edges to create a seamless repair.

d. Store the vehicle properly:
- When not in use, store the vehicle in a dry, covered area, such as a garage or carport, to protect it from rain, snow, and other moisture.
- If storing the vehicle for an extended period, consider using a breathable car cover or applying a rust-inhibiting wax or sealant to provide additional protection.

By treating bare metal surfaces after rust removal, applying appropriate rust-inhibiting coatings and treatments, following proper surface preparation techniques, and implementing regular maintenance and inspection practices, you can effectively prevent future rust development and keep your vehicle looking and performing its best for years to come. Remember, rust prevention is an ongoing process that requires diligence and attention to detail, but the effort is well worth it for the long-term health and value of your vehicle.

Painting and Sealing Repaired Areas

After treating bare metal surfaces and applying rust-inhibiting coatings, the final step in the rust repair process is painting and sealing the repaired areas to restore the vehicle's appearance and protect the newly treated metal from the elements. Properly painting and sealing repaired areas involves careful surface preparation, selection of appropriate paint products, and application of clear coats or sealants to provide a durable, long-lasting finish. In this section, we'll explore the key steps and best practices for painting and sealing rust repair areas, as well as tips for achieving a seamless, professional-looking result.

Surface Preparation for Painting
Proper surface preparation is essential for ensuring that the new paint adheres well to the repaired area and blends seamlessly with the surrounding finish. To prepare the surface for painting, follow these steps:

a. **Clean the surface:**
- Use a degreaser or solvent to remove any dirt, oil, or contaminants from the repaired area and surrounding paint.
- Pay close attention to seams, crevices, and other hard-to-reach areas where dirt and debris may accumulate.

b. **Sand the repaired area:**
- Use sandpaper in progressively finer grits (120-grit to 400-grit) to smooth and feather the edges of the repaired area, creating a gradual transition to the surrounding paint.

- Be careful not to sand too aggressively, as this can create deep scratches or expose bare metal.

c. **Apply body filler (if necessary):**
- If the repaired area has any minor imperfections or low spots, apply a thin layer of body filler using a flexible spreader or putty knife.
- Allow the filler to dry completely, then sand it smooth using progressively finer grits of sandpaper (80-grit to 400-grit).

d. **Prime the repaired area:**
- Apply a high-quality, rust-inhibiting primer to the repaired area using a spray gun or brush, following the manufacturer's instructions for application and drying times.
- Sand the primer smooth using fine-grit sandpaper (400-grit to 600-grit), then wipe away any dust or debris with a tack cloth.

Selecting and Applying Paint

Choosing the right paint products and application techniques is crucial for achieving a durable, professional-looking finish on rust repair areas. Consider the following factors when selecting and applying paint:

a. **Choose a compatible paint system:**
- Select a paint system that is compatible with the primer and substrate you're working with, as well as the surrounding paint on the vehicle.

- For most rust repair projects, a two-part, catalyzed paint system (base coat/clear coat) is recommended for optimal durability and appearance.

b. Match the paint color:
- Use the vehicle's paint code (usually found on a sticker or plate in the engine bay or door jamb) to obtain a custom-mixed paint that matches the original color.
- If the surrounding paint has faded or oxidized, consider blending the new paint into the adjacent panels to create a more seamless transition.

c. Apply the base coat:
- Using a spray gun, apply the base coat color to the repaired area in thin, even coats, allowing each coat to flash (become matte and dry to the touch) before applying the next.
- Build up the color gradually, applying additional coats as needed to achieve full coverage and opacity.

d. Apply the clear coat:
- Once the base coat has dried completely, apply a high-quality, compatible clear coat using a spray gun, following the manufacturer's instructions for application and flash times.
- Apply the clear coat in thin, even coats, allowing each coat to flash before applying the next.
- Apply a minimum of two to three coats of clear for optimal durability and gloss.

Blending and Finishing Techniques

To achieve a seamless, invisible repair, it's important to blend the new paint into the surrounding finish and refine the surface for a smooth, glossy appearance. Consider the following blending and finishing techniques:

a. Feather the edges:
- When sanding the repaired area and applying primer, be sure to feather the edges of the repair, gradually tapering the thickness of the primer and paint to create a smooth transition to the surrounding finish.
- Use a soft-edge foam sanding block or finely abraded blending disc to feather the edges and avoid creating hard lines or ridges.

b. Blend the paint:
- To help the new paint blend invisibly with the surrounding finish, use a blending technique, such as a "mist coat" or "fade out" method.
- For a mist coat, apply a light, dry coat of the base color over the repaired area and slightly beyond, helping to soften the transition between the new and old paint.
- For a fade out, gradually reduce the pressure on the spray gun trigger as you move away from the repaired area, allowing the new paint to gradually fade into the surrounding finish.

c. Color sand and polish:
- After the clear coat has cured completely (typically 24-48 hours), use a fine-grit sanding block (1500-grit to 3000-grit) to gently sand the surface, removing any minor imperfections or orange peel.

- Use a polishing compound and foam pad to buff the sanded area to a high gloss, restoring clarity and smoothness to the finish.
- If necessary, use a finer polishing compound or glaze to further refine the finish and remove any swirl marks or haze.

Sealing and Protecting the Repaired Area

To provide long-lasting protection and durability for the repaired area, consider applying a sealant or wax coating after painting and finishing. Here are some options for sealing and protecting rust repair areas:

a. Apply a paint sealant:
- Paint sealants are synthetic, polymer-based coatings that provide a durable, protective barrier over the painted surface.
- Apply the sealant according to the manufacturer's instructions, typically using a foam applicator pad or microfiber towel.
- Allow the sealant to cure for the recommended time before buffing to a high gloss.

b. Apply a ceramic coating:
- Ceramic coatings are advanced, nanotechnology-based coatings that bond chemically to the painted surface, providing exceptional hardness, durability, and resistance to UV rays, chemicals, and contamination.
- Professional application is recommended for ceramic coatings, as the process involves careful surface preparation and controlled application conditions.

- When properly applied and maintained, ceramic coatings can provide years of protection and enhanced gloss for rust repair areas and the entire vehicle.

c. Apply a carnauba wax:
- Carnauba wax is a natural, plant-based wax that provides a warm, deep shine and moderate protection for painted surfaces.
- Apply the wax using a foam applicator pad or microfiber towel, following the manufacturer's instructions for application and buffing.
- Reapply the wax every few months or as needed to maintain protection and shine.

By properly preparing the surface, selecting compatible paint products, using blending and finishing techniques, and applying protective coatings, you can achieve a seamless, professional-looking finish on rust repair areas that will last for years to come. Remember, the key to a successful rust repair is patience, attention to detail, and a commitment to following best practices at every step of the process. With the right techniques and products, you can restore your vehicle's appearance and protect it from future rust damage, ensuring that it looks and performs its best for miles to come

Chapter 7
Advanced Auto Body Repair Techniques
Working with Aluminum Body Panels

In recent years, the use of aluminum body panels in vehicle manufacturing has become increasingly common, as automakers seek to reduce weight, improve fuel efficiency, and enhance performance. However, working with aluminum requires specialized knowledge, tools, and techniques compared to traditional steel body repair. In this section, we'll explore the unique properties of aluminum, the challenges it presents in auto body repair, and the advanced techniques and best practices for working with aluminum body panels.

Properties of Aluminum
To effectively repair aluminum body panels, it's essential to understand the unique properties of this material and how they differ from steel. Key properties of aluminum include:

a. Lightweight:
- Aluminum has a density about one-third that of steel, making it significantly lighter for a given volume.
- This lighter weight contributes to improved fuel efficiency and performance in vehicles.

b. Strength and stiffness:
- While aluminum is not as strong as steel in terms of tensile strength, it has a better strength-to-weight ratio, meaning it can provide similar strength at a lower weight.

- Aluminum also has a lower modulus of elasticity than steel, meaning it is less stiff and more flexible.

c. Corrosion resistance:
- Aluminum naturally forms a thin, protective oxide layer when exposed to air, making it highly resistant to corrosion.
- However, aluminum can still be susceptible to galvanic corrosion when in contact with dissimilar metals, such as steel.

d. Thermal conductivity:
- Aluminum has a higher thermal conductivity than steel, meaning it dissipates heat more quickly.
- This property can affect welding and other heat-based repair techniques.

Challenges in Aluminum Body Repair
Working with aluminum body panels presents several unique challenges compared to traditional steel repair. These challenges include:

a. Softness and malleability:
- Aluminum is softer and more malleable than steel, meaning it can be more easily deformed or stretched during repair processes.
- This softness can make it challenging to restore the original shape of damaged panels and maintain proper body lines and contours.

b. Heat sensitivity:
- Aluminum has a lower melting point than steel and is more sensitive to heat during welding and other repair processes.
- Overheating aluminum can cause weakening, distortion, or even melting of the metal.

c. Corrosion concerns:
- While aluminum is naturally corrosion-resistant, it can still be susceptible to galvanic corrosion when in contact with other metals.
- Proper insulation and separation of dissimilar metals are crucial in aluminum repair to prevent corrosion.

d. Specialized tools and equipment:
- Repairing aluminum body panels requires specialized tools and equipment designed specifically for use with this material.
- Using tools or equipment intended for steel repair can contaminate or damage aluminum, compromising the integrity of the repair.

Aluminum Repair Techniques

To address the unique challenges of aluminum body repair, technicians must use advanced techniques and follow specific best practices. Key aluminum repair techniques include:

a. Dent repair:
- Aluminum dent repair often involves the use of specialized tools, such as aluminum-specific dent pulling systems and heat-based techniques, to reshape the metal without causing further damage.

- Paintless dent repair (PDR) techniques may also be used for minor dents, but technicians must be cautious not to overwork the softer aluminum.

b. Welding:
- Welding aluminum requires the use of specific alloys and techniques, such as tungsten inert gas (TIG) or metal inert gas (MIG) welding with aluminum-specific wire and shielding gas.
- Proper cleaning, joint preparation, and temperature control are critical to prevent contamination, distortion, or weakening of the aluminum during welding.

c. Riveting and bonding:
- In some cases, riveting and adhesive bonding may be used as alternatives to welding for joining aluminum panels.
- These techniques can help to avoid the heat-related challenges of welding and provide strong, durable joints.

d. Straightening and shaping:
- Aluminum's softness and malleability can make it challenging to straighten and shape damaged panels accurately.
- Technicians may use specialized aluminum straightening tools, such as precision hammers and dollies, along with heat-based techniques to gradually reshape the metal while minimizing the risk of stretching or thinning.

Best Practices for Aluminum Body Repair

To ensure the highest quality and durability of aluminum body repairs, technicians should follow these best practices:

a. Use dedicated tools and equipment:
- Always use tools and equipment specifically designed for aluminum repair, and keep them separate from those used for steel to prevent cross-contamination.
- This includes sanding and grinding tools, welding equipment, and even repair bay areas.

b. Ensure proper cleaning and preparation:
- Thoroughly clean aluminum surfaces before and during the repair process to remove any contaminants, such as dirt, oil, or oxidation.
- Use aluminum-specific cleaners and abrasives to avoid introducing foreign materials that could weaken or contaminate the metal.

c. Follow manufacturer guidelines:
- Consult the vehicle manufacturer's repair guidelines and specifications for the specific aluminum alloy and repair procedures required.
- Different aluminum alloys may have different properties and requirements, so it's crucial to follow the manufacturer's recommendations closely.

d. Maintain proper temperature control:
- Use a temperature-measuring device, such as an infrared thermometer, to monitor the temperature of the aluminum during welding or heat-based repair processes.

- Keep the temperature within the recommended range to prevent overheating, weakening, or distortion of the metal.

e. Protect against corrosion:
- When joining aluminum to dissimilar metals, use proper insulation and separation techniques, such as coatings, bushings, or spacers, to prevent galvanic corrosion.
- After the repair, apply corrosion-resistant primers and coatings specifically formulated for use on aluminum to provide long-term protection.

By understanding the unique properties of aluminum, mastering advanced repair techniques, and following best practices, technicians can successfully work with aluminum body panels and deliver high-quality, durable repairs. As the use of aluminum in vehicle manufacturing continues to grow, staying current with the latest techniques and technologies for aluminum repair will be increasingly important for auto body professionals.

Repairing Plastic Components

Plastic components have become increasingly common in modern vehicles, thanks to their lightweight, flexible, and cost-effective properties. From bumpers and fenders to interior trim and structural parts, plastics play a crucial role in automotive design and manufacturing. However, repairing damaged plastic components requires specialized knowledge and techniques that differ from traditional metal repair. In this section, we'll explore the types of automotive plastics, the challenges in plastic repair, and the advanced techniques and best practices for restoring damaged plastic parts.

Types of Automotive Plastics

Understanding the different types of plastics used in automotive applications is essential for selecting the appropriate repair approach. Common automotive plastics include:

a. Polypropylene (PP):
- PP is a versatile, lightweight, and cost-effective plastic commonly used for bumpers, fenders, and interior trim.
- It has good chemical resistance and can be easily molded into various shapes.

b. Polyethylene (PE):
- PE is a flexible, durable plastic used for fuel tanks, interior panels, and some bumper components.
- It has excellent impact resistance and can withstand low temperatures without becoming brittle.

c. Acrylonitrile Butadiene Styrene (ABS):
- ABS is a strong, rigid plastic used for interior trim, dashboards, and some exterior body parts.
- It has good dimensional stability and can be easily painted or chrome-plated.

d. Polyurethane (PU):
- PU is a versatile plastic used for bumpers, spoilers, and other exterior body parts.
- It has excellent flexibility, impact resistance, and can be easily molded into complex shapes.

e. Polycarbonate (PC):
- PC is a transparent, impact-resistant plastic used for headlight lenses, sunroofs, and some exterior trim parts.
- It has good heat resistance and dimensional stability, but can be prone to scratching.

Challenges in Plastic Repair

Repairing plastic components presents several unique challenges compared to metal repair, including:

a. Identification of plastic type:
- Different plastics have different chemical compositions and properties, which can affect the repair process.
- Identifying the specific type of plastic is crucial for selecting the appropriate repair materials and techniques.

b. Flexibility and memory:
- Many automotive plastics are designed to be flexible and resilient, which can make it challenging to restore their original shape after damage.

- Some plastics also have a "memory" effect, where they tend to return to their original shape after being deformed.

c. Heat sensitivity:
- Plastics have lower melting points than metals and can be easily damaged by excessive heat during repair processes.
- Careful temperature control and the use of specialized low-heat techniques are essential to prevent warping, discoloration, or weakening of the plastic.

d. Adhesion and compatibility:
- Some plastics may not bond well with certain adhesives, fillers, or paints, leading to poor repair quality and durability.
- Selecting compatible repair materials and following proper surface preparation techniques are critical for achieving strong, long-lasting repairs.

Plastic Repair Techniques

To address the challenges of plastic repair, technicians must use advanced techniques and specialized products. Key plastic repair techniques include:

a. Plastic welding:
- Plastic welding involves using heat to melt and fuse damaged plastic parts back together.
- Various welding methods can be used, such as hot air welding, ultrasonic welding, or nitrogen plastic welding, depending on the type of plastic and the extent of the damage.

- Proper technique and temperature control are essential to prevent overheating or distortion of the plastic.

b. Adhesive bonding:
- Adhesive bonding involves using specialized plastic adhesives to join damaged parts or fill in gaps and cracks.
- The success of adhesive bonding depends on proper surface preparation, including cleaning, sanding, and applying plastic adhesion promoters.
- Technicians must select adhesives that are compatible with the specific type of plastic being repaired.

c. Plastic filler and putty:
- Plastic fillers and putties are used to fill in small holes, cracks, or imperfections in damaged plastic parts.
- These products are typically two-part epoxy or polyester-based and can be sanded, shaped, and painted once cured.
- Proper mixing, application, and curing techniques are crucial for achieving a strong, seamless repair.

d. Reinforcing and backing:
- For larger areas of damage or structural repairs, reinforcing mesh or backing materials may be used to provide additional strength and stability.
- These materials can be adhered to the backside of the damaged area before applying filler or adhesive to help bridge gaps and distribute stress.

Best Practices for Plastic Repair

To ensure the highest quality and durability of plastic repairs, technicians should follow these best practices:

a. Identify the plastic type:
- Use identification methods, such as part markings, density tests, or chemical tests, to determine the specific type of plastic being repaired.
- Consult the vehicle manufacturer's repair guidelines and specifications for the recommended repair techniques and materials.

b. Clean and prepare surfaces:
- Thoroughly clean the damaged area and surrounding surfaces to remove any dirt, grease, or contaminants that could interfere with the repair process.
- Use plastic-specific cleaners and abrasives to avoid damaging or weakening the plastic.
- Apply plastic adhesion promoters or primers as recommended by the manufacturer to ensure proper bonding of repair materials.

c. Control temperature:
- Use low-heat repair techniques and tools, such as plastic welders with adjustable temperature settings, to avoid overheating or distorting the plastic.
- Monitor the temperature of the plastic during the repair process, keeping it within the recommended range for the specific type of plastic.

d. Select compatible materials:
- Choose repair materials, such as adhesives, fillers, and paints, that are specifically formulated for use on the type of plastic being repaired.
- Follow the manufacturer's guidelines for mixing, application, and curing of repair materials to ensure proper performance and durability.

e. Reinforce and protect:
- Use reinforcing mesh or backing materials as needed to provide additional strength and stability for larger repairs or structural components.
- After the repair is complete, apply plastic-specific primers, paints, and clear coats to protect the repaired area from UV damage, chemicals, and other environmental factors.

By mastering advanced plastic repair techniques and following best practices, technicians can successfully restore damaged plastic components, improving the appearance, functionality, and value of the vehicle. As automotive plastics continue to evolve and expand in use, staying up-to-date with the latest repair technologies and techniques will be essential for auto body professionals.

Fixing Dents on Difficult Locations

Dent repair is a common task in auto body work, but some dents can be particularly challenging due to their location, size, or the complexity of the panel shape. Difficult-to-repair dents may be found in areas with limited access, tight corners, or on panels with multiple curves or contours. In this section, we'll explore the challenges of fixing dents in difficult locations and discuss advanced techniques and tools for addressing these complex repairs.

Challenges of Difficult Dent Locations

Dents in challenging locations present several obstacles that can make the repair process more complex and time-consuming, such as:

a. Limited access:
- Some dents may be located in areas that are hard to reach from behind, such as near the edges of panels, in tight corners, or close to structural components.
- Limited access can make it difficult to use traditional dent repair tools and techniques, such as hammers, dollies, or stud welders.

b. Complex panel shapes:
- Dents on panels with multiple curves, contours, or body lines can be challenging to repair without distorting the original shape of the panel.
- Maintaining the precise geometry of the panel while removing the dent requires a high level of skill and specialized techniques.

c. Stretched metal:
- In some cases, the impact that caused the dent may have stretched the metal beyond its original shape, making it difficult to restore without causing further damage.
- Stretched metal may require the use of heat, shrinking, or other advanced techniques to gradually restore the panel to its pre-damage condition.

d. Adjacent components:
- Dents located near other vehicle components, such as lights, trim pieces, or mechanical parts, may require additional disassembly or protection to avoid causing collateral damage during the repair process.
- Removing and reinstalling these components can add significant time and complexity to the repair.

Advanced Dent Repair Techniques for Difficult Locations

To successfully repair dents in challenging locations, technicians may need to employ advanced techniques and specialized tools. Some of these techniques include:

a. Paintless Dent Repair (PDR):
- PDR is a non-invasive method that involves using specialized tools to massage the dent out from behind the panel, without disturbing the paint.
- For difficult-to-reach areas, technicians may use long-reach tools, such as telescoping rods or bendable handles, to access the backside of the dent.
- PDR can be particularly effective for smaller dents on panels with limited access, as it minimizes the need for disassembly and refinishing.

b. Glue pulling:
- Glue pulling involves using specialized adhesives and tabs to pull the dent out from the front side of the panel.
- This technique can be useful for dents in areas where access to the backside of the panel is limited, or where PDR tools may not be able to apply sufficient pressure.
- Technicians must be careful to use the appropriate adhesive strength and pulling force to avoid causing further damage to the paint or panel.

c. Shrinking and stretching:
- For dents that have stretched the metal, a combination of shrinking and stretching techniques may be necessary to restore the panel's original shape.
- Shrinking involves applying controlled heat to the stretched area, causing the metal to contract and return to its pre-stretched state.
- Stretching involves using a stud welder or other tool to gently pull the metal back into shape, gradually smoothing out the dent.

d. Precision hammering and dollying:
- For dents in complex panel shapes or tight corners, precision hammering and dollying techniques may be required to restore the contours and body lines.
- This involves using specialized hammers and dollies with various shapes and curves to carefully tap the dent out from both sides of the panel, while maintaining the correct geometry.
- Technicians must have a high level of skill and patience to avoid over-correcting the dent or creating new distortions in the panel.

Specialized Tools for Difficult Dent Repair

In addition to advanced techniques, repairing dents in difficult locations may require the use of specialized tools designed for specific challenges. Some of these tools include:

a. Long-reach PDR tools:
- Telescoping rods, bendable handles, and other long-reach tools allow technicians to access dents in tight spaces or hard-to-reach areas.
- These tools often feature interchangeable tips with various shapes and sizes to accommodate different dent configurations and panel contours.

b. Glue pulling kits:
- Glue pulling kits include a variety of tabs, bridges, and pulling devices designed for different dent sizes and locations.
- These kits also include specialized adhesives with varying strength levels to ensure a secure bond without damaging the paint.

c. Induction heaters:
- Induction heaters use electromagnetic fields to quickly and precisely heat specific areas of a panel for shrinking or stress relief.
- This allows technicians to apply controlled heat to stretched metal without risking overheating or damaging surrounding areas.

d. Precision hammers and dollies:
- Precision hammers and dollies come in a wide variety of shapes, sizes, and curves to match the contours of different panel types and dent locations.
- These tools are often made of high-quality materials, such as forged steel or tungsten, to provide the necessary strength and durability for precise dent shaping.

Best Practices for Repairing Dents in Difficult Locations

When addressing dents in challenging areas, technicians should follow these best practices to ensure the best possible repair outcome:

a. Assess the dent carefully:
- Before beginning the repair, thoroughly inspect the dent and surrounding area to identify any complicating factors, such as stretched metal, limited access, or adjacent components.
- Use a bright light source and a straight edge to evaluate the depth, size, and shape of the dent accurately.

b. Choose the appropriate repair method:
- Based on the assessment, select the repair method (PDR, glue pulling, shrinking, etc.) that is best suited for the specific dent location and characteristics.
- Consider factors such as accessibility, panel shape, and potential for paint damage when choosing the repair approach.

c. Protect surrounding areas:
- Use masking tape, foam pads, or other protective materials to shield adjacent panels, trim pieces, or mechanical components from accidental damage during the repair process.
- Take care to avoid applying excessive pressure or heat that could transfer to nearby areas and cause unintended distortion or damage.

d. Work gradually and patiently:
- Dent repair in difficult locations often requires a gradual, iterative approach to avoid overcorrecting or creating new problems.
- Use gentle, controlled movements with the appropriate tools and techniques, and check progress frequently to ensure the dent is being removed evenly and accurately.

e. Verify the repair quality:
- After completing the dent removal, carefully inspect the repaired area under different lighting conditions to check for any remaining imperfections, distortions, or unevenness.
- Use a straight edge or body line tape to confirm that the panel contours and geometry have been restored to their original condition.

By understanding the challenges of dent repair in difficult locations, employing advanced techniques and specialized tools, and following best practices, technicians can successfully restore even the most complex dents to a high standard of quality.

Ongoing training and practice in these advanced repair methods will help technicians build the skills and confidence needed to tackle any dent, regardless of its location or severity.

Custom Modifications and Bodywork

Custom modifications and bodywork involve altering the appearance or functionality of a vehicle beyond its original factory specifications. These modifications can range from subtle cosmetic changes to extensive structural and mechanical alterations, allowing vehicle owners to express their individual style, improve performance, or adapt their vehicles for specific purposes. In this section, we'll explore the various types of custom modifications and bodywork, the skills and techniques required to execute them, and the best practices for ensuring safe, legal, and high-quality results.

Types of Custom Modifications and Bodywork
Custom modifications and bodywork can take many forms, depending on the vehicle owner's goals, preferences, and budget. Some common types of modifications include:

a. Body kits and aerodynamic enhancements:
- Installing aftermarket body kits, spoilers, splitters, or diffusers to change the vehicle's appearance and improve aerodynamic performance.
- These modifications can be made from various materials, such as fiberglass, carbon fiber, or polyurethane, and may require custom fitting and painting to blend seamlessly with the vehicle's existing body.

b. Custom paint and graphics:
- Applying unique paint colors, patterns, or graphics to the vehicle's body to create a one-of-a-kind appearance.
- This can involve anything from a simple color change to intricate airbrushed designs, vinyl wraps, or hand-painted murals.

c. **Widebody conversions and fender flares:**
- Modifying the vehicle's body to accommodate wider wheels and tires, often for improved handling or a more aggressive stance.
- This can involve installing aftermarket fender flares, wide body kits, or custom fabricating new body panels to achieve the desired width and shape.

d. **Chassis and suspension modifications:**
- Altering the vehicle's chassis or suspension components to improve handling, stability, or ride height.
- This can include installing lowering springs, coilovers, sway bars, or custom fabricating new suspension mounting points or subframes.

e. **Interior modifications:**
- Customizing the vehicle's interior with aftermarket seats, steering wheels, gauges, or audio/video systems.
- This can also involve modifying the interior structure, such as removing or relocating seats, creating custom consoles or enclosures, or installing roll cages or harness bars for safety or performance.

Skills and Techniques for Custom Bodywork

Executing custom modifications and bodywork requires a diverse set of skills and techniques, depending on the specific project and the materials involved. Some essential skills include:

a. **Fabrication and welding:**
  - Creating custom body panels, brackets, or structural components from raw materials, such as sheet metal, tubing, or composite fabrics.
  - Using various welding techniques, such as MIG, TIG, or oxyacetylene, to join metal components securely and cleanly.

b. **Fiberglass and composite work:**
  - Working with fiberglass, carbon fiber, or other composite materials to create custom body parts, such as hoods, fenders, or spoilers.
  - This involves laying up the composite fabric, applying resin, and shaping the part over a mold or form until it cures to the desired shape and strength.

c. **Bodywork and surface preparation:**
  - Preparing the vehicle's body for custom modifications by repairing any existing damage, filling and sanding imperfections, and ensuring a smooth, even surface for paint or graphics.
  - This may involve using body fillers, primers, and various sanding and shaping tools to achieve the desired contours and finish quality.

d. **Custom painting and finishing:**
  - Applying custom paint colors, graphics, or finishes to the modified body components using advanced techniques, such as color matching, blending, masking, and clear coating.

- This requires a keen eye for detail, steady hand, and knowledge of paint chemistry and application methods to achieve a flawless, durable finish.

e. **Mechanical and electrical integration:**
- Integrating the custom body modifications with the vehicle's existing mechanical and electrical systems, such as engine, transmission, suspension, or lighting.
- This may require custom fabrication of mounting points, wiring harnesses, or control modules to ensure proper fit, function, and safety.

Best Practices for Custom Modifications and Bodywork

When planning and executing custom modifications and bodywork, it's essential to follow best practices to ensure the safety, legality, and quality of the finished project. Some key considerations include:

a. **Compliance with regulations and standards:**
- Research and adhere to all applicable laws, regulations, and safety standards related to vehicle modifications, such as bumper heights, lighting requirements, or structural integrity.
- Ensure that any custom modifications do not compromise the vehicle's crashworthiness, visibility, or emissions compliance.

b. **Proper planning and design:**
- Carefully plan and design the custom modifications before beginning any work, considering factors such as material selection, structural integrity, fitment, and aesthetics.

- Use computer-aided design (CAD) software, 3D modeling, or physical prototypes to visualize and refine the design before committing to final fabrication.

c. **High-quality materials and craftsmanship:**
- Use high-quality, durable materials that are appropriate for the specific application and can withstand the stresses and environmental conditions of automotive use.
- Execute all fabrication, welding, and finishing work with a high level of skill, precision, and attention to detail to ensure a professional-grade result.

d. **Testing and quality control:**
- Thoroughly test and inspect all custom modifications for proper fit, function, and safety before final installation or use.
- Conduct road tests, stress tests, or other quality control measures to verify that the modifications perform as intended and do not adversely affect the vehicle's handling, braking, or other critical systems.

e. **Documentation and communication:**
- Keep detailed records and documentation of all custom work performed, including design sketches, material specifications, and step-by-step processes.
- Communicate clearly with the vehicle owner or client about the scope, timeline, and cost of the project, as well as any potential risks, limitations, or maintenance requirements associated with the modifications.

Chapter 8
Troubleshooting Common Auto Body Repair Issues
Paint Imperfections and How to Fix Them

A flawless paint finish is a critical component of any high-quality auto body repair job. However, achieving a perfect paint finish requires skill, attention to detail, and the ability to troubleshoot and correct various types of paint imperfections. In this section, we'll examine common paint imperfections that can occur during the repair process, explore their causes, and provide step-by-step guidance on how to fix them for a smooth, glossy, and durable finish.

Types of Paint Imperfections
Paint imperfections can manifest in various forms, each with its own distinct characteristics and underlying causes. Some common types of paint imperfections include:

a. Orange peel:
- Orange peel refers to a textured, dimpled appearance on the paint surface, resembling the skin of an orange.
- This imperfection is often caused by improper spray technique, incorrect gun setup, or inadequate paint flow and leveling.

b. Runs and sags:
- Runs and sags appear as thick, droopy accumulations of paint, usually on vertical surfaces or in corners and crevices.

- They are typically caused by applying too much paint in one area, using an incorrect spray technique, or not allowing sufficient flash time between coats.

c. **Pinhole pops:**
- Pinhole pops are tiny, crater-like defects that appear on the surface of the paint, often in clusters.
- They can be caused by contaminants in the paint or on the surface, trapped air bubbles, or improper paint preparation and application.

d. **Solvent popping:**
- Solvent popping occurs when small bubbles or blisters form on the paint surface, usually during the drying or curing process.
- This imperfection is often caused by trapped solvents or moisture in the paint layers, or by applying paint in excessively thick coats.

e. **Peeling and lifting:**
- Peeling and lifting refer to the separation or detachment of the paint film from the underlying surface or between paint layers.
- These issues can be caused by improper surface preparation, contamination, incompatible paint products, or extreme environmental conditions.

Causes of Paint Imperfections

To effectively troubleshoot and prevent paint imperfections, it's essential to understand the underlying causes that contribute to their formation. Some common causes of paint imperfections include:

a. **Improper surface preparation:**
- Failing to properly clean, sand, or treat the surface before painting can lead to adhesion issues, contamination, or uneven paint application.
- Skipping crucial steps, such as priming or sealing, can also contribute to paint defects and failures.

b. **Incorrect paint mixing or application:**
- Using the wrong mixing ratios, incompatible reducers or hardeners, or expired paint products can cause a variety of paint imperfections.
- Applying paint in excessively heavy or thin coats, or not allowing sufficient flash time between coats, can also lead to defects like runs, sags, or solvent popping.

c. **Equipment issues and setup:**
- Using damaged, dirty, or improperly maintained spray equipment can result in poor atomization, uneven paint flow, or contamination.
- Incorrect gun settings, such as air pressure, fluid tip size, or fan pattern, can also contribute to paint imperfections.

d. **Environmental factors:**
- Extreme temperatures, humidity, or airborne contaminants in the painting environment can interfere with paint application, drying, and curing processes.
- Painting in direct sunlight, windy conditions, or poorly ventilated areas can also lead to various paint defects.

Fixing Paint Imperfections

Once paint imperfections have been identified and their causes have been determined, the next step is to take corrective action to fix the issues and restore a smooth, flawless finish. The specific approach to fixing paint imperfections will depend on the type and severity of the defect, but some general steps include:

a. Sanding and leveling:
- For imperfections like orange peel, runs, or sags, the first step is to sand down the affected area to level the paint surface.
- Use progressively finer grits of sandpaper (1000 to 3000 grit) to remove the defects and create a smooth, even base for refinishing.
- Be careful not to sand through the paint layers or into the underlying substrate.

b. Cleaning and preparation:
- After sanding, thoroughly clean the area to remove any dust, debris, or contaminants that could interfere with the repair process.
- Use a tack cloth, compressed air, or a solvent-based cleaner to ensure a clean, dry surface for refinishing.
- If necessary, apply a primer or sealer to promote adhesion and prevent further imperfections.

c. Spot repair and blending:
- For small, localized imperfections, such as pinhole pops or solvent popping, a spot repair approach may be sufficient.

- Apply a small amount of paint or clear coat to the affected area, using a fine-tipped brush or a small spray gun.
- Blend the repair into the surrounding paint to create a seamless transition, using techniques like a "mist coat" or a "fade-out" process.

d. Full panel refinishing:
- For more extensive or severe paint imperfections, a full panel refinishing may be necessary to achieve a consistent, high-quality result.
- Sand and level the entire panel, then clean and prepare the surface for painting.
- Apply the appropriate primer, base coat, and clear coat layers, following the manufacturer's guidelines for mixing, application, and flash times.
- Use proper blending techniques to ensure a smooth transition between the refinished panel and the adjacent areas.

Best Practices for Preventing Paint Imperfections

In addition to knowing how to fix paint imperfections, it's equally important to take proactive steps to prevent them from occurring in the first place. Some best practices for preventing paint imperfections include:

a. Thorough surface preparation:
- Always follow the proper steps for cleaning, sanding, and treating the surface before painting.
- Use high-quality abrasives, cleaners, and primers that are compatible with the paint system and the substrate.

- Allow adequate drying and curing time between each step to ensure proper adhesion and performance.

b. Proper paint mixing and application:
- Follow the manufacturer's guidelines for mixing ratios, viscosity, and pot life of the paint products.
- Use the appropriate reducers, hardeners, and additives for the specific paint system and environmental conditions.
- Apply the paint in thin, even coats, allowing sufficient flash time between each layer to prevent runs, sags, or solvent popping.

c. Maintaining a clean, controlled environment:
- Keep the painting area clean, organized, and free from dust, dirt, and other contaminants.
- Control the temperature, humidity, and ventilation in the painting booth to ensure optimal drying and curing conditions.
- Use high-quality, properly maintained air filtration and exhaust systems to minimize airborne particles and overspray.

d. Regular equipment maintenance and calibration:
- Clean and maintain all spray guns, hoses, and other painting equipment according to the manufacturer's recommendations.
- Check and calibrate the equipment regularly to ensure proper fluid flow, atomization, and fan pattern.
- Replace worn or damaged parts, such as fluid tips or air caps, to prevent inconsistent or defective paint application.

By understanding the causes of paint imperfections, mastering the techniques for fixing them, and implementing best practices for prevention, auto body technicians can achieve consistently high-quality paint finishes that enhance the appearance, durability, and value of the repaired vehicle. Ongoing training, attention to detail, and a commitment to continuous improvement will be essential for staying at the forefront of this critical aspect of auto body repair.

Uneven or Misaligned Body Panels

Stubborn dents can be a frustrating and time-consuming challenge for auto body repair technicians. These dents may resist conventional repair techniques due to their location, size, depth, or the characteristics of the metal involved. In this section, we'll explore the factors that contribute to stubborn dents, the advanced techniques and tools used to address them, and best practices for achieving successful, long-lasting repairs.

Factors Contributing to Stubborn Dents
Several factors can make a dent more difficult to repair, including:

a. Location:
- Dents located in hard-to-reach areas, such as near the edges of panels, in tight corners, or close to body lines or creases, can be challenging to access with conventional dent repair tools.
- Dents on panels with complex curves, angles, or reinforcements may also resist straightening or pulling due to the inherent rigidity of the structure.

b. Size and depth:
- Large, deep dents may require more extensive repair techniques or even panel replacement, as the metal may have stretched or deformed beyond its ability to be straightened.
- Shallow, wide dents can also be stubborn, as they may not provide enough leverage or purchase for dent pulling tools to work effectively.

c. **Metal characteristics:**
- The type, thickness, and temper of the metal can affect its ability to be straightened or reformed.
- High-strength steel, aluminum, or other advanced alloys used in modern vehicles may require specialized repair techniques or heat treatment to avoid cracking, tearing, or weakening the metal.

d. **Previous repairs:**
- Dents that have been previously repaired or filled may be more challenging to address, as the existing filler or reinforcements can interfere with the straightening process.
- In some cases, the previous repair may need to be completely removed and the area stripped back to bare metal before attempting to repair the stubborn dent.

Advanced Techniques for Stubborn Dent Repair

To successfully repair stubborn dents, auto body technicians may need to employ advanced techniques and specialized tools beyond the standard hammer and dolly or paintless dent repair (PDR) methods. Some of these techniques include:

a. **Metal shrinking:**
- For dents that have stretched or deformed the metal, shrinking the affected area with controlled heat application can help restore its original shape.
- This technique involves heating the metal with a torch or induction heater to a specific temperature range, then cooling it rapidly to shrink the metal fibers.

- Proper temperature control and technique are crucial to avoid overheating, burning, or weakening the metal.

b. Stud welding:
- Stud welding involves attaching small metal studs or pins to the surface of the dent, then using a slide hammer or pulling bridge to apply gradual, controlled pressure to pull the dent out.
- This technique can be effective for larger, deeper dents or those in hard-to-reach areas where conventional PDR tools may not have sufficient access or leverage.
- Care must be taken to avoid overheating or damaging the metal during the welding process, and the studs must be properly removed and the area cleaned and finished after pulling.

c. Partial panel replacement:
- In some cases, the most effective solution for a stubborn dent may be to replace a portion of the affected panel, rather than attempting to repair it.
- This technique involves cutting out the damaged section of the panel and welding in a new, pre-formed replacement piece that matches the original contours and shape.
- Proper alignment, welding, and finishing techniques are essential to ensure a seamless, invisible repair that maintains the structural integrity of the panel.

d. Pressure reshaping:
- For stubborn dents on panels with complex curves or angles, pressure reshaping techniques may be used to gradually reform the metal to its original shape.

- This can involve the use of specialized clamps, levers, or hydraulic tools to apply controlled pressure to the backside of the dent while supporting the surrounding metal.
- Patience and finesse are key to this technique, as rushing or applying too much pressure can cause further distortion or damage to the metal.

Specialized Tools for Stubborn Dent Repair

In addition to advanced techniques, repairing stubborn dents may require the use of specialized tools designed for specific challenges or applications. Some examples include:

a. **Dent pulling systems:**
- Advanced dent pulling systems, such as stud welders, slide hammers, or glue pullers, can provide the necessary leverage and control to remove stubborn dents.
- These systems often feature interchangeable attachments, adjustable bridges, or multi-point pulling mechanisms to adapt to different dent shapes and locations.

b. **Induction heaters:**
- Induction heaters use electromagnetic fields to quickly and precisely heat metal surfaces for shrinking or stress relieving.
- These tools allow for targeted heating of specific areas without the risk of overheating or damaging surrounding materials.

c. **Specialized hammers and dollies:**
- A wide range of specialized hammer and dolly shapes, sizes, and materials are available for tackling stubborn dents in various locations and panel contours.
- These tools may feature unique profiles, such as curved or angled faces, or be made from lightweight, non-marring materials to prevent additional damage to the metal.

d. **Pressure reshaping tools:**
- Hydraulic or pneumatic pressure reshaping tools, such as clamps, rams, or levers, can apply controlled force to reform stubborn dents gradually.
- These tools often feature adjustable pressure settings, interchangeable tips, or pivoting mechanisms to adapt to different panel shapes and access angles.

Best Practices for Stubborn Dent Repair

When dealing with stubborn dents, auto body technicians should keep the following best practices in mind to ensure successful, long-lasting repairs:

a. **Assess the dent carefully:**
- Take the time to thoroughly examine the dent and surrounding area, noting any complicating factors such as metal stretching, previous repairs, or limited access.
- Consider the type, thickness, and temper of the metal involved, as well as any potential challenges posed by the panel shape or structure.
- Use this assessment to select the most appropriate repair technique and tools for the specific situation.

b. **Work gradually and patiently:**
- Stubborn dents often require a gradual, multi-step approach to avoid overcorrecting or causing additional damage.
- Apply pressure or heat slowly and incrementally, checking progress frequently and making adjustments as needed.
- Allow adequate cooling or settling time between steps to prevent distortion or stresses in the metal.

c. **Protect surrounding areas:**
- When using heat, welding, or pressure techniques, take care to protect adjacent panels, trim, or sensitive components from damage.
- Use heat shields, welding blankets, or masking materials to isolate the repair area and prevent collateral damage.

d. **Maintain proper technique:**
- Follow proper safety procedures and use appropriate personal protective equipment (PPE) when working with heat, welding, or pressure tools.
- Maintain consistent technique and tool control to ensure even, predictable results and minimize the risk of mistakes or accidents.
- Keep tools and equipment properly maintained and calibrated to ensure optimal performance and results.

e. **Verify the repair quality:**
- After completing the repair, thoroughly inspect the area to ensure that the dent has been fully removed and the panel has been restored to its original shape and contour.

- Check for any signs of distortion, thinning, or weakening of the metal that may compromise the structural integrity or appearance of the repair.
- Test the repair under different lighting conditions and viewing angles to confirm a seamless, invisible result.

By understanding the unique challenges posed by stubborn dents, employing advanced repair techniques and specialized tools, and following best practices for assessment, execution, and quality control, auto body technicians can successfully tackle even the most difficult dent repairs. Continuously developing skills and knowledge in this area will be essential for staying competitive and delivering high-quality results in a demanding industry.

Preventing and Fixing Rust Recurrence

Rust recurrence is a common and frustrating problem in auto body repair, as it can compromise the integrity, appearance, and value of a vehicle even after extensive repair efforts. Preventing and fixing rust recurrence requires a thorough understanding of the causes and contributing factors, as well as the implementation of effective strategies and techniques to address the underlying issues. In this section, we'll explore the key concepts and best practices for preventing and fixing rust recurrence in auto body repair.

Understanding Rust Recurrence

Rust recurrence occurs when rust reappears on a previously repaired or treated surface, often in the same location or pattern as the original rust damage. Several factors can contribute to rust recurrence, including:

a. Inadequate surface preparation:
- Failing to properly clean, strip, or treat the metal surface before applying rust repair or prevention products can allow rust to continue forming beneath the surface.
- Skipping steps such as sanding, grinding, or chemical treatment can leave behind rust residue or scale that can promote future corrosion.

b. Incomplete rust removal:
- If rust is not completely removed from the affected area, including any pits, crevices, or hard-to-reach spots, it can continue to spread and cause recurrence.
- Using inadequate tools, techniques, or products for rust removal can also leave behind traces of rust that can later resurface.

c. Improper product selection or application:
- Using rust repair or prevention products that are not suitable for the specific type of metal, environment, or application can result in poor performance and recurrence.
- Applying products incorrectly, such as in the wrong thickness, sequence, or curing conditions, can also compromise their effectiveness and longevity.

d. Environmental factors:
- Exposure to moisture, salt, chemicals, or other corrosive elements can accelerate rust recurrence, especially if the repaired area is not adequately sealed or protected.
- Storing or operating the vehicle in humid, coastal, or industrial environments can also increase the risk of rust recurrence.

Strategies for Preventing Rust Recurrence

To prevent rust recurrence, auto body technicians should employ a comprehensive, multi-faceted approach that addresses the root causes and risk factors. Some key strategies include:

a. Thorough surface preparation:
- Always follow proper procedures for cleaning, stripping, and treating the metal surface before beginning any rust repair or prevention work.
- Use appropriate abrasives, cleaners, and chemicals to remove all traces of rust, scale, and contaminants, including in hard-to-reach areas.

- Ensure that the surface is completely dry and free of debris before applying any coatings or treatments.

b. Complete rust removal:
- Use a combination of mechanical and chemical methods to remove all visible and hidden rust from the affected area, including pits, crevices, and seams.
- Consider using specialized tools, such as abrasive blasters, needle scalers, or chemical rust removers, to access and eliminate hard-to-reach rust deposits.
- Verify that all rust has been removed by visually inspecting the surface and testing with a magnet or chemical indicator.

c. Proper product selection and application:
- Choose rust repair and prevention products that are specifically formulated for the type of metal, environment, and application involved.
- Follow the manufacturer's instructions carefully for mixing, applying, and curing the products, ensuring that the correct thickness, sequence, and conditions are used.
- Use high-quality, compatible primers, fillers, and topcoats to provide additional layers of protection and durability.

d. Sealing and protecting the repaired area:
- After completing the rust repair and refinishing process, apply a high-quality, durable sealant or coating to the repaired area to prevent moisture and contaminants from penetrating.

- Consider using additional protection methods, such as cavity wax, undercoating, or anti-corrosion films, in high-risk areas or environments.
- Ensure that all seams, joints, and edges are properly sealed and protected to prevent rust from spreading or recurring.

Techniques for Fixing Rust Recurrence
If rust recurrence does occur despite prevention efforts, auto body technicians must take prompt and thorough action to address the issue and prevent further damage. Some effective techniques for fixing rust recurrence include:

a. Identifying and assessing the recurrence:
- Carefully inspect the affected area to determine the extent and severity of the rust recurrence, noting any changes or progression from the original damage.
- Look for signs of underlying causes, such as moisture intrusion, product failure, or mechanical damage, that may have contributed to the recurrence.
- Document the recurrence with detailed notes, photographs, and measurements to guide the repair process and communicate with customers or insurance providers.

b. Removing the rust and affected materials:
- Completely remove all rust and any compromised repair materials, such as failed fillers, primers, or coatings, from the affected area.
- Use a combination of abrasive, chemical, and mechanical methods to ensure that all rust and debris are eliminated, including in pits, crevices, and seams.

- Take care to avoid damaging or removing excess metal or adjacent components during the removal process.

c. **Treating and preparing the surface:**
- After removing the rust and affected materials, thoroughly clean and treat the exposed metal surface to prevent further corrosion and promote adhesion of repair products.
- Use appropriate cleaners, abrasives, and chemical treatments, such as phosphoric acid or conversion coatings, to etch and stabilize the metal surface.
- Ensure that the surface is completely dry and free of contaminants before proceeding with the repair process.

d. **Applying rust repair and prevention products:**
- Select high-quality, compatible rust repair and prevention products that are appropriate for the specific metal, environment, and application.
- Apply the products in the correct sequence, thickness, and curing conditions, following the manufacturer's instructions carefully.
- Use a combination of primers, fillers, and topcoats to provide multiple layers of protection and ensure a durable, long-lasting repair.

e. **Verifying and monitoring the repair:**
- After completing the rust recurrence repair, thoroughly inspect the area to ensure that all rust has been eliminated and the repair products have been applied correctly.

- Test the adhesion, hardness, and compatibility of the repair materials to verify that they will provide adequate protection and performance.
- Monitor the repaired area over time, especially in high-risk environments or conditions, to detect any signs of recurrence or failure early.

Best Practices for Rust Recurrence Prevention and Repair
To maximize the effectiveness and longevity of rust recurrence prevention and repair efforts, auto body technicians should follow these best practices:

a. **Educate customers on rust prevention:**
- Provide customers with information and guidance on how to prevent rust formation and recurrence on their vehicles, including regular washing, waxing, and inspections.
- Recommend appropriate rust prevention products, such as cavity wax or undercoating, for customers in high-risk environments or with rust-prone vehicles.
- Encourage customers to address any signs of rust or damage promptly to prevent more extensive and costly repairs down the line.

b. **Use high-quality, compatible products:**
- Invest in premium-grade rust repair and prevention products that are specifically formulated for the types of metals, environments, and applications encountered in auto body repair.

- Ensure that all products used in the repair process, including abrasives, cleaners, primers, fillers, and topcoats, are compatible with each other and with the substrate metal.
- Follow the manufacturer's recommendations for product selection, mixing, application, and curing to ensure optimal performance and results.

c. **Document and communicate the repair process:**
- Keep detailed records of all rust recurrence prevention and repair work, including the specific products, techniques, and steps used in each case.
- Use this documentation to communicate with customers, insurance providers, and other stakeholders about the scope, quality, and effectiveness of the repair work.
- Maintain open lines of communication with customers to address any concerns, questions, or feedback they may have about the repair process or results.

d. **Continuously improve skills and knowledge:**
- Stay up-to-date with the latest products, techniques, and best practices for rust recurrence prevention and repair through ongoing training, education, and professional development.
- Participate in industry events, workshops, and certification programs to expand skills and knowledge in this critical area of auto body repair.
- Share knowledge and experiences with colleagues and peers to foster a culture of continuous improvement and excellence in rust recurrence prevention and repair.

By understanding the causes and contributing factors of rust recurrence, implementing effective prevention and repair strategies, and following best practices for product selection, application, and documentation, auto body technicians can successfully address this common and challenging issue. Providing customers with high-quality, long-lasting rust repairs not only enhances the value and integrity of their vehicles but also helps to build trust, loyalty, and referrals in a competitive industry.

Conclusion

Congratulations on making it to the end of this comprehensive guide on auto body repair! By now, you should have a solid understanding of the various techniques, tools, and best practices involved in repairing automotive dents, fixing bumpers, and eliminating hail damage.

Throughout this book, we've covered a wide range of topics, from the fundamentals of paintless dent repair and traditional dent removal methods to advanced techniques for working with aluminum, plastic, and other challenging materials. We've also explored the intricacies of rust repair, paint matching, and panel alignment, providing step-by-step guidance and expert tips for achieving professional-quality results.

But beyond just the technical skills and knowledge, we've also emphasized the importance of developing a customer-centric approach to auto body repair. By prioritizing clear communication, attention to detail, and a commitment to continuous improvement, you can build a reputation for excellence and establish long-lasting relationships with your clients.

As you move forward in your auto body repair journey, remember to stay curious, adaptable, and open to new ideas and techniques. The industry is constantly evolving, with new materials, technologies, and repair methods emerging all the time. By staying up-to-date with these developments and continuously refining your skills, you'll be well-positioned to tackle any challenge that comes your way.

We hope that this book has not only provided you with practical guidance and insights but also inspired you to pursue your passion for auto body repair with renewed energy and enthusiasm. Whether you're a seasoned professional or just starting out in the field, always remember the value and importance of your work in helping people restore their cherished vehicles to their former glory.

Thank you for joining us on this educational journey, and we wish you all the best in your future endeavors as an auto body repair technician. Keep learning, keep growing, and keep making a difference, one dent at a time!

www.ingramcontent.com/pod-product-compliance
Lightning Source LLC
Chambersburg PA
CBHW050057230526
45470CB00004B/1565